Aerospace Power

A Brazilian Multidimensional Approach

EDITING COMMITTEE:
PEDRO ARTHUR LINHARES LIMA
CARLOS EDUARDO VALLE ROSA
CARLOS ALBERTO LEITE DA SILVA
EDUARDO OLIVEIRA SOL DA SILVA
GILLS VILAR LOPES
Translated by MARISA HELENA DE OLIVEIRA SILVA

Air University Press
Maxwell Air Force Base, Alabama

Air University Press

Director
Dr. Paul Hoffman

Managing Editor
Dr. Christopher Rein

Design and Production Managing Editor
Luetwinder T. Eaves

Project Editor
Donna Budjenska

Project Facilitator
Jorge F. Serafin, Lt Col (Ret.), USAF

Contributing Editors
Michael Labosky
Kimberly Leifer

Cover Art, Book Design, and Illustrations
Catherine Smith

Composition and Prepress Production
Cheryl Ferrell

Air University Press
600 Chennault Circle, Building 1405
Maxwell AFB, AL 36112-6010
https://www.airuniversity.af.edu/AUPress/

Facebook:
https://facebook.com/AirUnivPress

Twitter:
https://twitter.com/aupress

LinkedIn:
https://www.linkedin.com/company/air-university-press/

Instagram:
https://www.instagram.com/air_university_press/

Library of Congress Cataloging-in-Publication Data

Names: Linhares, Pedro Arthur, author. | Air University (U.S.). Press, issuing body.
Title: Aerospace power: a Brazilian multidimensional approach / Pedro Arthur Linhares [and four others].
Description: Description: Maxwell Air Force Base, Alabama : Air University Press, [2023] | In scope of the U.S. Government Publishing Office Cataloging and Indexing Program (C&I); Federal Depository Library Program (FDLP) distribution status to be determined upon publication. | Includes bibliographical references. | Summary: "This collection of studies delves into themes of Brazil's airpower and space power: protecting territorial sovereignty in the Amazon, developing missile technology in partnership with South Africa, taking a place in global aerospace geopolitics, participating in UN peace operations, among many more. Subject matter experts and officers from the Força Aerea Brasileira (FAB) provide in-depth research not found anywhere else but this anthology"—Provided by publisher.
Identifiers: LCCN 2022057955 (print) | LCCN 2022057956 (ebook) | ISBN 9781585663255 (paperback) | ISBN 9781585663262 (Adobe PDF)
Subjects: LCSH: Air power--Brazil. | Brazil. Fôrça Aérea Brasileira. | Brazil—Military relations. | Aerospace industries—Brazil.
Classification: LCC UG635.B7 L56 2023 (print) | LCC UG635.B7 (ebook) | DDC 358.400981—dc23/eng/20230104 | SUDOC D 301.26/6:B 73
LC record available at https://lccn.loc.gov/2022057955
Published by Air University Press in March 2023

Disclaimer

Opinions, conclusions, and recommendations expressed or implied within are solely those of the authors and do not necessarily represent the official policy or position of the organizations with which they are associated or the views of the Air University Press, Air University, United States Air Force, Department of Defense, or any other US government agency. This publication is cleared for public release and unlimited distribution.

Reproduction and printing are subject to the Copyright Act of 1976 and applicable treaties of the United States. This document and trademark(s) contained herein are protected by law. This publication is provided for noncommercial use only. The author has granted nonexclusive royalty-free license for distribution to Air University Press and retains all other rights granted under 17 U.S.C. §106. Any reproduction of this document requires the permission of the author.

This book and other Air University Press publications are available electronically at the AU Press website: https://www.airuniversity.af.edu/AUPress.

Unless otherwise indicated, figures obtained via the US Department of Defense. The appearance of US Department of Defense (DOD) visual information does not imply or constitute DOD endorsement.

Contents

List of Illustrations	*v*
Foreword	*vii*
About the Authors	*ix*
Introduction	*xiii*
Carlos Eduardo Valle Rosa	

Part 1: Aerospace Power Applications

Introduction to Part 1	3
1 Small Remotely Piloted Aircraft Weaponization	5
Leland Delgado Assis	
Afonso Farias de Sousa Júnior	
2 Cyber Defense in the Brazilian Air Force	19
Pedro Arthur Linhares Lima	
Gills Vilar Lopes	
André Lucas Alcântara da Silva	
Ines Correa Gomes Cardinot	
3 SIPAM and SIVAM Projects	27
Luiz Fernando Póvoas da Silva	
Flavio N. H. Jasper	
4 Brazilian Air Traffic Flow Management	43
Eduardo Sol Oliveira da Silva	
Joaquim Lobo Júnior	

Part 2: Aerospace Power and Contemporary Issues

Introduction to Part 2	59
5 Brazil and International Humanitarian Law	63
Luciano Vaz Ferreira	
Carlos Alberto Leite da Silva	
Luís Eduardo Pombo Celles Cordeiro	
6 The Brazilian Air Force in UN Peace Operations	79
Pedro Henrique Nascimento dos Santos	
Claudia Maria Sousa Antunes	

CONTENTS

7   Interoperability among Brazil's Armed Forces   93
    Marta Maria Telles
    Alessandra Veríssimo Lima Santos
    Rainer Ferraz Passos

8   Geopolitics, Culture, and Law   103
    Guilherme Sandoval Góes
    Maria Célia Barbosa Reis da Silva

9   Aerospace Geopolitics   115
    Carlos Eduardo Valle Rosa

Part 3: Aerospace Logistics and Economics

Introduction to Part 3   131

10 Logistics Principles and the Axioms of Combat   135
    Luiz Tirre Freire
    Fábio Ayres Cardoso

11 Offset Practice as a Public Policy   153
    Alexander de Mello Lima
    Rodrigo Antônio Silveira dos Santos

12 Transfer of Technology in Military Projects   169
    Carlos Roberto Santos
    Patrícia de Oliveira Matos

Conclusion   193
Carlos Eduardo Valle Rosa

Abbreviations   197

Bibliography   205

Index   237

Illustrations

Figures

3.1. SIPAM and SIVAM Concept ... 32

3.2. SISDACTA's Concept ... 34

3.3. Evolution of SISDACTA ... 35

12.1. The joint participation of Brazil (COMAER) and South Africa (RSA DoD) ... 178

12.2. A-Darter missile subsystems and national companies ... 180

Tables

6.1. UN Peacekeeping Capability Readiness System ... 85

6.2. Brazilian participation with aircraft in UN peace operations ... 86

10.1. Results of the semantic analysis (axioms of combat) ... 146

10.2. Results of the semantic analysis (principles of logistics) ... 147

12.1. Dimensions of the Contingent Effectiveness Model ... 172

12.2. Technology transfer effectiveness assessment model synthesis ... 173

Foreword

Brazil is among the largest countries in the world and the largest country in Latin America, covering an expanse of over 8.5 million square kilometers. It borders every country in South America except Ecuador and Chile and contains within its borders the resource-rich and increasingly ecologically significant Amazon River basin and rainforest. Brazil has an estimated population approaching 215 million people, which makes it one of the most populous countries on the planet. Indeed, it is estimated that approximately one third of the population of Latin America resides in Brazil. In short, Brazil is one of the most important countries in the Western Hemisphere and, indeed, globally.

The Brazilian military—especially the Brazilian air force—continues to play a vital role, not only in protecting and promoting Brazilian national interests, but also, more broadly, in hemispheric defense. Indeed, the Força Aérea Brasileira (FAB), founded in 1941, is the largest air force in the Southern Hemisphere and the second largest in the Western Hemisphere. This volume, *Aerospace Power: A Brazilian Multidimensional Approach*, represents a timely and poignant analysis and assessment of Brazil's aerospace defense industries, its aerospace resources, and its effectiveness in applying those resources in response to contemporary challenges and opportunities. In doing so, it builds upon its 2017 companion volume, *Culture and Defence in Brazil: An Inside Look at Brazil's Aerospace Strategies*, which highlighted Brazil's security dilemmas and the ways in which the FAB's professional military educational system prepares airmen to respond to these contemporary challenges and opportunities.

The publication of this important book through the USAF Air University Press is a symbol of Brazilian-American friendship in several ways. It is emblematic of the deep collegial friendship between the US Air Force and the FAB—a friendship that is a central component to Inter-American cooperation and hemispheric defense. It also symbolizes the mutually valuable, synergistic relationship between the USAF Air University and Universidade da Força Aérea (UNIFA, the Brazilian air force university) and, more specifically, the present and future collaborative relationship between UNIFA and Air University Press. This synergistic partnership serves to mutually strengthen both US and Brazilian professional military educational institutions.

I am confident that this volume, combined with its earlier 2017 companion volume, will serve to greatly enhance our knowledge and appre-

FOREWORD

ciation of the vitally important role played by the FAB, which, in turn, constitutes one of the US's principal partners in defending and building a better future throughout the Western Hemisphere and globally.

> HOWARD M. HENSEL, PHD
> Henry H. Arnold Professor of
> International Relations and
> Strategic Studies
> USAF Air War College

About the Authors

Claudia Maria Sousa Antunes, PhD, is a professor and researcher at Brazil's Universidade da Força Aérea (UNIFA, Air Force University) Postgraduate Program and author of "Mulher combatente: políticas e marcações de gênero na defesa"[Combatant women: gender policies and markings in defense].

Leland Delgado Assis, PhD candidate, is an instructor, Brazilian Air Force Command and Staff College, and a lieutenant colonel in the Força Aérea Brasileira (FAB, Brazilian air force).

Ines Correa Gomes Cardinot, MSc candidate, is a researcher in the Pro-Defense IV, researcher at the Simulations and Scenarios Laboratory at the Escola de Guerra Naval (Naval Warfare Academy), and author of "A Crisis within a Crisis: Impact on Critical Public Administration Information Infrastructure during the Pandemic."

Fábio Ayres Cardoso, PhD, is a professor, Air Force University Postgraduate Program, and a colonel in the FAB as well as chief of Field Logistics Group and author of "Resources Package Modelling Supporting Border Surveillance Operations."

Luís Eduardo Pombo Celles Cordeiro, MSc, is an instructor and member of the United Nations Integrated Multidimensional Mission for the Stabilization of the Central African Republic and a lieutenant colonel in the FAB.

Luciano Vaz Ferreira, PhD, is a professor at the Federal University of Rio Grande, Brazil.

Luiz Tirre Freire, PhD, is a professor at the Air Force University Postgraduate Program, the Brazilian Air Force Command and Staff College, the School of Army Command and Staff College, the Superior School of War, and the School of Socio-Educational Management. Freire also authored "Impact on the Defense Budget," commanded the Logistics Command of the operations area Ágata VI Operation, and is a retired brigadier, FAB.

Guilherme Sandoval Góes, Post-PhD, is a professor, Air Force University Postgraduate Program, professor (emeritus) at the Army Command and Staff College, and a retired captain, Brazilian Navy, as well as author of "The Geopolitics of Energy in the 21st Century."

ABOUT THE AUTHORS

Flavio Neri Hadmann Jasper, PhD, is a professor at the Air Force University Postgraduate Program and a retired colonel, FAB, as well as author of "The Brazilian Model of Air Space Control: Culture and Defence in Brazil: An Inside Look at Brazil's Aerospace Strategies."

Alexander de Mello Lima, MSc candidate (MBA in Strategic Management, MBA in Public Administration), is an electronic engineer and a major, FAB.

Pedro Arthur Linhares Lima, PhD, is a professor at the Air Force University Postgraduate Program, production engineer, coordinator of Strategic Studies Center at Air Force University, a researcher at Rede CTIDC/Pró-Defesa IV (CAPES/MD), and a retired brigadier, FAB.

Joaquim Lôbo Júnior, MSc, is rapporteur for the AFIS Subgroup of the International Civil Aviation Organization and head of the Air Traffic Flow Management Operational Hall of the Air Navigation Management Center.

Patrícia de Oliveira Matos, PhD, is a professor, Air Force University Postgraduate Program, and economist as well as author of "Industrial Base Defense: An Analysis of the Space Systems Industry in Brazil."

Rainer Ferraz Passos, MSc, is an electronic and computer engineer; a colonel, FAB; as well as author of "A Quantitative and Qualitative Approach to Targeting as a Weapon-Target Assignment."

Carlos Eduardo Valle Rosa, PhD, is a professor at Air Force University Postgraduate Program; a retired colonel, FAB; and author of "Poder Aéreo: Guia de Estudos" [Air Power: A Guide to Studies].

Alessandra Veríssimo Lima Santos, MSc, is an independent researcher and author of "Formação Militar: reflexões sobre a abordagem da Inteligência Cultural" [Military Training: Reflections on the Cultural Intelligence Approach].

Carlos Roberto Santos, MSc, is a manager at NAV Brazil, a retired FAB Reserve officer, and author of "The A-Darter Binational Project and its Technology Transfer Model."

Pedro Henrique Nascimento dos Santos, MSc candidate, is an International Operations Coordinator and author of "Brazil in Peace Operations and the South American Contribution for Peace and International Security."

ABOUT THE AUTHORS

Rodrigo Antônio Silveira dos Santos, PhD, is a professor at the Air Force University Postgraduate Program and part of the Control and Monitoring Group at SAAB AB in Linköping-Sweden, working on the F-X2 (Gripen) Project as well as a colonel, FAB.

André Lucas Alcântara da Silva, MSc candidate, is a computer engineer and a lieutenant, FAB, serving as chief of the Cyber Defense Cell on Aerospace Operations Command.

Carlos Alberto Leite da Silva, MSc, PhD, is a professor at the Air Force University Postgraduate Program and dean of Extension and Cooperation at the University of the Air Force.

Eduardo Sol Oliveira da Silva, PhD, is a professor at the Air Force University Postgraduate Program and vice coordinator of the Postgraduate Program at the University of the Air Force.

Luiz Fernando Póvoas da Silva, PhD, is a professor at the Air Force University Postgraduate Program and a retired colonel, FAB.

Maria Célia Barbosa Reis da Silva, PhD, is a professor at the Air Force University Postgraduate Program and the Brazilian Superior School of War as well as editor, Brazilian Superior School of War Magazine.

Afonso Farias de Sousa Júnior, PhD, is a professor at the Air Force University Postgraduate Program; a retired colonel, FAB; and author of the chapter "Human Dignity and Sovereignty: Analysis of the Constitutionality of the Destruction Shot," in Culture and Defense in Brazil.

Marta Maria Telles, PhD, is a collaborating lecturer at the Air Force University Postgraduate Program and author of "Armed Forces Integration: The Brazilian Case."

Gills Vilar Lopes, PhD, is a professor and program coordinator at the Air Force University Postgraduate Program as well as a researcher at Rede CTIDC/Pró-Defesa IV (CAPES/MD).

Introduction

Carlos Eduardo Valle Rosa

Welcome, readers! To provide context for those new to Brazilian aerospace power, the first lesson is that, in Brazil, the following elements comprise such power:

a) the Brazilian air force, the element that fundamentally aggregates military capability; b) civil aviation, whose strength represents—besides the potential reserve of human resources and materials in case of mobilization—the economic and geopolitical importance of reaching regions that are difficult to access; c) aerospace infrastructure, an essential support factor to aviation activities due to the size of Brazil; d) the defense and aerospace industry, in accordance with the demand for products and services of international quality; e) aerospace technological and scientific complex, which ensures technological autonomy and generates cutting-edge knowledge in the aviation and space fields; and f) scientists and researchers, with expertise in aerospace activity, a requirement for the development of technological-scientific solutions for challenges of a field in which training is prime.[1]

Background

The Brazilian government created the Air Force Ministry on 20 January 1941, gathering the army's and navy's aviation resources. The Civil Aviation Department was subordinated to the Ministry of Roads and Public Works. At that time, the National Air Mail (Correio Aéreo Nacional) was also created, unifying the extant naval and military airmail. The Brazilian air force (Força Aérea Brasileira, FAB) was created 22 May 1941. Brazilian aviation grew episodically and in circumstances such as popular insurgencies and separatist movements in the country.

However, this scenario changed in World War II. The nascent FAB fought the Italian-German submarine threat in the South Atlantic and collaborated with Allied war efforts in Italy. Both experiences granted Brazilian pilots a different perspective of airpower. Flying cutting-edge airships, employing modern tactics with state-of-the-art equipment, comprehending the relevance of the aviation industry in supporting and developing the best means, and observing the dependency on airport infrastructure awakened a new consciousness regarding the meaning of airpower apart from the experience of deploying airplanes in combat.

The FAB's first basic doctrine, published in 1958, stated that airpower goes beyond combat activities. This doctrine is an official document of great internal importance, as it it leads the thinking on the institution's main focus. Therefore, airpower would have constituting elements like the "Air Force, whose components are its air bases, reserves, installations and supporting military organizations; Civil Aviation; the aviation infrastructure, which includes the airport network, the flight protection and the air traffic control systems; the Aviation Industry; and the aviation affairs research institutes."[2]

This trend of incorporating various constituting elements in the concept of airpower was not limited to basic doctrine; the Ministry of Aviation also introduced several initiatives. One of these concepts was cited by Nelson Freire Lavenère-Wanderley, who observed that the creation of an "Aviation Technical Center ... for the formation of engineers, and a research center"[3] for the development of space and aviation technology had been studied since 1946. The consolidation of this center initiated the main aviation industry in Latin America: Embraer (now a world leader in commercial aircraft).

The Brazilian Airspace Control System (Sistema de Controle do Espaço Aéreo Brasileiro) pioneered the idea of uniting civil air traffic control and air defense, and the FAB is currently still responsible for coordinating civil and military air traffic within the 22 million square kilometers under Brazilian jurisdiction—including land and maritime areas defined in international agreements.

The air force was also in charge of building airports, especially in the Amazon region, making it an organization with deep knowledge of airport engineering, capable of handling the tropical rainforest challenges. These airports soon became a network whose objective was to integrate the isolated region into the rest of the country. This geopolitical perspective, led by men such as Brigadier General Lysias Rodrigues, led to the National Air Mail, responsible for bringing basic state services, such as essential health, education, citizenship, and supplies, to populations isolated and reachable only by airplanes, especially in the Amazon rainforest.

In 1975, a new version of the FAB's basic doctrine introduced the concept of aerospace power, advancing views first noted in doctrine published in 1958.[4] This 1975 doctrine expanded the atmospheric geography to outer space. The Cold War, the space race, and Brazil joining select nations with space programs through its air force all influenced the expansion of airpower views to aerospace power.

From the time of aerostats to airplanes and now to satellites, a natural progression from "air" to "space" and, consequently, "aerospace" evolved. That also occurs from a geographic perspective when the extension of one domain/environment/dimension toward the other takes place. From a technological view, the integration of capabilities is a trend observed in the applications of civil and military nature. In the field of economy, the aerospace industry demonstrates the interrelation among industry activities. At the symbolic level, or what is typically referred to as soft power, both aviation and astronautics spark attention to matters related to prestige and power.

The idea of aerospace power, representing human and material, technological, infrastructural, economic, civil, and military capabilities—both in aviation and space arenas—is reality in Brazil. This is true also in public policy; the 2020 *National Defense Policy* (*Política Nacional de Defesa, PND*) and the *National Defense Strategy* (*Estratégia Nacional de Defesa, END*)[5] assign the Air Force the responsibility for space sector activities, in addition to continuing its sovereignty in airspace. To bolster this responsibility and attendant capability, budgeting among the Brazilian armed forces, which does not spawn challenges as intense as in other nations, enjoys relative stability.

Organization

Thus, an aerospace power volume that gathers contributions from Brazilian scholars must be characterized by this comprehensive, multidimensional awareness stretching back to the creation of the Brazilian air force. Texts from faculty and students of the Air Force University (Universidade da Força Aérea, UNIFA) Post-Graduation Program in Aerospace Sciences (Programa de Pós-Graduação em Ciências Aeroespaciais, PPGCA) include coverage of the economy and technological development, logistics, humanitarian rights issues in peace operations, interoperability aspects in joint operations, themes of the

INTRODUCTION

aircraft's military capabilities employment and in the cyber context, air traffic control and geopolitics, and culture.

This book is divided in three main parts, which concentrate similar themes. This organization seeks to reinforce the understanding of a comprehensive and multifaceted vision of aerospace power, as it is understood in Brazil.

Part 1, Aerospace Power Applications, consists of chapters whose themes are related to aircraft, resilience to cyberattacks, surveillance capability, command and control (C2), and military and civilian air traffic control. The chapters explore the multidimensional aerospace power concept, in which technological evolution plays a fundamental role and which guides the demands of sovereignty and national and regional development. In "Small Remotely Piloted Aircraft Weaponization: A Simulated Study Compared to Embraer A-29 Capabilities," Leland Delgado Assis and Afonso Farias de Sousa Júnior use system theory to conclude that small systems of remotely piloted aircraft (RPA), in the context of fourth-generation wars with different actors from state entities, may have their use strengthened by the social impact of their deployment. That allows a better understanding of risk analysis in military planning using RPAs in the analyzed scenarios. Continuing in the context of military applications, Pedro Arthur Linhares Lima, Gills Vilar Lopes, André Lucas Alcântara da Silva, and Ines Correa Gomes Cardinot discuss cyberspace and the FAB's responsibility in "Cyber Defense in the Brazilian Air Force: Overview and Future Perspectives." The authors analyze the Air Force Cybernetic Defense Center's role as part of the FAB, revealing its structure and importance as it faces cyber threats in the aerospace power context, including in aerial platforms.

The FAB's systems for air traffic control and monitoring in the Amazon, the Amazon Protection System (Sistema Integrado de Proteção da Amazônia, SIPAM) and the Amazon Surveillance System (Sistema de Vigilância da Amazônia, SIVAM), come under scrutiny by Luiz Fernando Póvoas da Silva and Flávio Neri Hadmann Jasper in "SIPAM and SIVAM Projects: A Challenge in the Amazon." These systems protect Brazil's borders and by extension its ecosystems and population in addition to providing the information necessary to maintain national sovereignty in the Amazon and its strategic surroundings. The main challenge is incorporating the idea of C2 in multidomain scenarios and in a modern hybrid war environment.

INTRODUCTION

SIPAM and SIVAM—in a setup uncommon to the rest of the world's air forces—also conduct all air traffic control for Brazil. The air traffic management (ATM) service is intrinsically connected to the FAB, which is responsible for both civil air traffic and aerospace defense. Eduardo Sol Oliveira da Silva and Joaquim Lobo Júnior reveal that this commitment goes beyond the national airspace in "Brazilian Air Traffic Flow Management: An Integration Link in South America." Through cooperation agreements, Brazil has been collaborating with other South American countries over the years in order to provide and support safer and more efficient air traffic control systems.

After reviewing these constituent elements of aerospace power, the text then connects to the themes discussed in Part 2, Aerospace Power and Contemporary Issues. The intersection of air and space power with current issues seems to generate expectations both domestically and internationally.

One of the issues is the role played by the United Nations (UN) in peacekeeping operations. The vision of common commitments established by international organizations reinforces expectations about the operations of armed forces. Studying the aerospace perspective in UN peacekeeping operations from an aerospace perspective underscores the demand for strict adherence to the legal mechanisms of international law. Aerospace power has been a tool of first choice for governments and international entities in solving security problems in places where the UN observes a lack of political and social stability. However, these expectations are not always met. Analyzing the origins of unmet expectations may point to another current issue: The performance of the armed forces, whether in peacekeeping operations or in their main defense role, reveals the importance of studies related to the preparation and use of aerospace power, especially in Brazil, where there are still difficulties in the interoperability issue.

Allied to this view of common commitments established by international bodies in armed forces' operations, Luciano Vaz Ferreira, Carlos Alberto Leite da Silva, and Luís Eduardo Pombo Celles Cordeiro focus on international humanitarian law (IHL) for national strategic issues, especially with regard to aerospace power, in "Brazil and International Humanitarian Law: Application in Strategic Sectors for National Defense." Brazilian defense documents are in accordance with the international legal precepts in force, providing legal

INTRODUCTION

basis for military actions outside the national territory and strengthening the vision of cooperation defended by the country in multilateral courts.

In accordance with the application of IHL and based on the United Nations indicators, such as Peacekeeping Capability Readiness System (UNPCRS), Pedro Henrique Nascimento dos Santos and Claudia Maria Sousa Antunes analyze FAB preparations to accomplish peacekeeping operations in "The Brazilian Air Force in UN Peace Operations: Ready to Support Global Stability." They also discuss the search for peace and the contribution to human rights and international standards to which armed forces engaged in this type of context must be submitted.

Themes such as interoperability are vital for integration in military operations. "Interoperability among the Brazilian Armed Forces: The Cultural Intelligence Perspective," by Marta Maria Telles, Alessandra Veríssimo Lima Santos, and Rainer Ferraz Passos, delves into joint operations from a cultural perspective. As Brazil continues implementing jointness as a standard of conduct, studies like these provide even broader approaches for understanding the multidimensionality of aerospace power.

Guilherme Sandoval Góes and Maria Célia Barbosa Reis da Silva next establish a dialog among geopolitics, culture, and law, deducing the effect on aerospace power. In "Geopolitics, Culture, and Law: Epistemological Convergences that Impact Aerospace Power," the authors consider geopolitical tensions of the capitalist triad inspired by John Locke and in the multipolar world reflected in Immanuel Kant. They further examine issues connected to the use of outer space in light of the great post–COVID-19 challenge of considering aerospace power in a multipolar, metaconstitutional, and multicivilizational world.

This approach fosters discussion on a wider range of aerospace power concepts that are purely national and puts geopolitics and international relations into perspective. In "Aerospace Geopolitics," Carlos Eduardo Valle Rosa shows that the aerospace environment is defined as a geographical domain formed by the combination of airspace and outer space in which geopolitical relations are established. From there, the geopolitical relevance of this environment is analyzed through geographical evidence and political, economic, technological, and ideological variables. The author considers the aerospace environment a new domain for geopolitical science and a stage for national defense and development strategies.

In Part 3, Aerospace Logistics and Economics, the connection between the constituent elements of aerospace power and the aerospace and defense industries, as well as with the aerospace scientific and technological complex, becomes clear. Ongoing concern with logistical issues—often associated with developing capabilities in the national aerospace industry—is evident in a country with limited financial resources.

Therefore, Part 3 returns to the initial postulate of the comprehensiveness and multidimensionality of aerospace power in Brazil. It highlights that logistics, aimed at sustaining the application of force, and economics, aimed at generating force, are integrated with aerospace power in Brazilian aerospace strategic thinking.

In the context of greater economy, logistics is important in military applications of aerospace power. The lack of recent combat experience requires that theoretical concepts support both military employment and combat experience practices. Fabio Ayres Cardoso and Luiz Tirre Freire seek to validate logistics principles in the FAB, in "Logistics Principles and the Axioms of Combat: An Analysis," based on combat and semantics theory. Their analysis uses combat axioms as a conceptual system to allow the construction of plausible solutions related to a set of concepts accepted as valid.

In the field of aerospace economy, Alexander de Mello Lima and Rodrigo Antônio Silveira dos Santos analyze the offset practices made possible by FAB processes regarding international purchases. Indeed, this practice has become a public policy, as demonstrated in "Offset Practice as Public Policy: The Case of the Brazilian Air Force." This chapter unveils offset as a national industry qualification measure, generating opportunities to better provide products and services that have a high value, which is characteristic of the aerospace sector. Therefore, the authors compare overall defense strategic goals and their legal basis with the FAB's offset practices, identifying opportunities for establishing offset agreements within a public policy to provide new capabilities for Brazil's aerospace industry.

The book concludes with a chapter from Carlos Roberto Santos and Patrícia de Oliveira Matos, who study strategic partnerships in the technological development of aerospace weapon systems. In "Transfer of Technology in Military Projects: A Case Study of the A-Darter Project," the authors highlight how important multilateral

INTRODUCTION

agreements are for the development of military capabilities, as was the case involving Brazil and South Africa with the A-Darter air-to-air missile, a highly complex technological system that, besides generating lessons for future projects, achieved the established technical requirements and enabled a product with operating characteristics among the best in the market.

Summary

Focusing on a variety of topics connected to current challenges in the aerospace domain, especially those in which Brazil seeks to act effectively at the international level, this compendium chronicles Brazil's evolution from airpower to aerospace power. Therefore, it is imperative that discussions on themes related not only to aerospace power but also to outer space and cybernetic space should take place. All these geographic domains become more and more relevant for humanity.

Also in accordance with the Brazilian view of aerospace power, the tridimensional perception would not be solely geographic but also theoretical, situational, and conceptual. Since the beginning of Brazilian aerospace strategic thinking dating to the 1950s, the great strategic challenge has been to understand aerospace power from an all-encompassing perspective, and this challenge is successfully answered and contextualized in the chapters of this work. This book brings important contributions in areas including military capability, national deployment concepts, national integration, support activities infrastructure, aerospace economy and industry, science and technology, and aviation culture.

Happy reading!

INTRODUCTION

Notes

(Notes appear primarily in shortened form. For full details, see the appropriate entry in the bibliography.)

1. Ministry of Defense (MOD), *DCA 1-1, DOUTRINA BÁSICA DA FORÇA AÉREA BRASILEIRA*, 28–30.
2. Ministry of Aeronautics, "Decree 1000 - GM2 and 'Doutrina Básica Da Força Aérea.'"
3. Lavenère-Wanderley, *Historia Da Força Aerea Brasileira*, 314.
4. Air Force Chief of Staff, *DMA 1-1 Doutrina Básica Da Força Aérea Brasileira*.
5. MOD, *Livro Branco de Defesa Nacional*.

Part 1
Aerospace Power Applications

Introduction to Part 1

Applications are the means by which aerospace power is employed (in combat, for example, but also in peacetime situations); these applications or tools derive from systems or technologies. The chapters of part 1 explore these current tools and related issues from a Brazilian context, bringing national perspectives on remotely piloted aircraft (RPA), cybersecurity, and air traffic control, whose dual characteristic (civilian and military integration) is highlighted in chapters 3 and 4.

The applications of aerospace power reveal the daily practice of specific functions. One of the essential functions of this power is its armed use, aiming kinetic effects against the opponent. A blunt instrument of this application that has proven to be efficient (including in international training such as the US Air Force Green Flag exercise) is the Embraer 314 (A-29 Super Tucano). By proposing a comparison of this aircraft with another important battlefield tool, the armed drone, the first chapter examines the realignment of these applications in view of the capabilities demonstrated by RPA. To this end, a comparative study on the effectiveness of aircraft, such as the A-29 in its employment vector role, with armed RPA is desirable. This is a hypothesis that highlights the importance of drones, whose presence on the battlefield is proliferating. For this reason, studies such as this are relevant to Brazilian aerospace power, whose final applications must follow contemporary technological trends.

This assertion is also true in the case of cybersecurity. The cyber domain is already characterized as a field of military confrontation, and states without cyber defense capabilities will be vulnerable to the actions of their eventual opponents. Importantly for Brazilian aerospace power, the cyber dimension is considered a new, fifth domain for operational capabilities, with a double role: either as an intermediate activity—supporting networks and information systems—or as a main activity, impacting doctrine. To this end, a proposed Brazilian air force cyber defense center could plan, coordinate, and centralize cybersecurity actions. Operationally, it would be responsible for guiding and executing protective measures, exploitation, and mitigation of damage resulting from cyberattacks, operating as a prompt response to network incidents, and supervising integrated actions.

This systemic vision of cyber protection already resides in Brazilian air traffic management, even since its creation. Brazil integrates aerospace defense with air traffic control in combined structures responsible for providing sovereignty in airspace and ensuring safe air movement over areas of national responsibility. For this reason, both activities are considered relevant applications of Brazilian aerospace power. In Northern Brazil—a region with monumental dimensions and geographic characteristics encompassing dense forest cover, high humidity, a vast hydrographic network, and more—the Amazon Surveillance System and the Amazon Protection System reflect this hybridism, as they are intended for aerospace defense and air traffic control, respectively. Both are designed to address challenges in the strategic Amazon environment, a multidomain environment. As relevant applications, the systems use sensors and available means to identify information in the infosphere, which, after analysis, can contribute to operational intelligence. This intelligence then can be harnessed in the construction of scenarios used as simulations that help decision-makers. This information is also used to support civil aviation and manage air traffic control.

Unlike in other states, Brazil's combined air defense and air movement control generates responsibilities and opportunities for greater integration with regional neighbors, especially in civil air traffic management. The development of aviation on a global scale has brought common demands to states (regulation, for example). But it also brings challenges, as in the case of sovereign transgressions. The International Civil Aviation Organization (ICAO) recommends regional partners work in an integrated and cooperative manner in Brazil. In an extensive area such as the one under Brazilian responsibility—covering 22 million square kilometers (8.5 million square miles)—the development of common structures, systems, and rules awakens a great national potential to influence the surrounding region. Therefore, Brazil, through support and integration programs in several neighboring countries, considers the air traffic management service a relevant application of aerospace power.

The concept of Brazilian aerospace power is multidimensional, and the following chapters highlight some key ideas of this concept: first, the representation of the technological evolution that underlies the aerospace environment; next, the integration of capabilities that provide adequate solutions to the demands of national and regional sovereignty and development; finally, the certainty that the applications of aerospace power, discussed in this first part, precisely reflect one segment of the Brazilian aerospace strategic thinking.

Chapter 1

Small Remotely Piloted Aircraft Weaponization
A Simulated Study Compared to Embraer A-29 Capabilities

Leland Delgado Assis
Afonso Farias de Sousa Júnior

Introduction

This chapter originated as part of a master's degree dissertation in the postgraduate program in Air and Space Sciences at the Brazilian air force university (Universidade da Força Aérea, UNIFA) and focuses on understanding the weaponized use of small remotely piloted aircraft (SRPA), also called small drones. We proposed a simulation and an analysis methodology comparing the employment of a variable number of SRPA with the employment of a fixed-wing aircraft in the Brazilian air force (Força Aérea Brasileira, FAB) inventory as a control group. We used the Embraer EMB-314 Super Tucano A-29 air-to-ground attack profile for this scenario. Designed with the characteristics of a light attack aircraft, it is ideal for operations in which the aerial vector is used to support friendly troops, necessarily in contact with the enemy, featuring so-called close-air support missions.[1] A brief examination of the evolution of armed conflicts will demonstrate the applicability of this research, which discusses the doctrinal aspects related to air and space control tasks linked to the guarantee and denial of the military employment of both domains.[2]

The end of the Thirty Years' War, in addition to rewriting the map of Western Europe, left an indelible mark on the collective consciousness of those involved due to a previously unseen level of destruction. Thus, the Treaty of Westphalia in 1648 "founded the modern state, by affirming the primacy of sovereigns in secular interests"[3] and consolidated the nation-state as a political entity with the exclusive claim to the legitimate use of violence, defining this period as First Generation War.[4]

Between 1648 and 1789, armies and navies professionalized and improved. The period from 1789 to 1914 was characterized by the influence of the Industrial Revolution.[5] In addition, an intense process of industrialization provoked the exponential improvement of the use of violence by states and sovereigns, with war being supported by mobilizing all national resources. Clausewitz's work discussed those improvements, used by politicians and military chiefs to think about war as a political and military phenomenon of politics by other means.[6] The vertiginous progress achieved by military engineering led to the development of equipment that further expanded war's destructive potential: machine guns, assault rifles, battleships, heavy weapons, submarines, and airplanes. For this reason, the First World War, with its predominance of heavy artillery, was the apex of Second Generation War.[7]

As societies progressed, the globalization of conflicts emerged; the Second World War was emblematic of this development, characterized by great mobility that defined Third Generation War.[8]

Subsequently, a prolonged Cold War resulting from the geopolitical disputes between the United States of America and the Soviet Union spun off several secondary conflicts, such as the Vietnam War and the Soviet invasion of Afghanistan.[9]

The end of the Cold War "seemed to underline the West's military, economic, and ideological dominance."[10] However, as Miranda and Nascimento stated, "the illusion created by the hope of a peaceful post–Cold War world has not materialized, and the world is increasingly armed, and conflicts multiplied. Along with the power to inflict material damage and deaths, in a new scenario of borderless war, the strategic consequence of the irregular warfare actions is destabilizing the state and imposing one ideology by force and fear."[11]

This context of irregular warfare without boundaries defines Fourth Generation War, in which the state has lost its violence monopoly when fighting nonstate opponents. As an aggravating factor, "practically everywhere the State is losing."[12]

Technological development is so significant that it characterizes the milestones or changes between generations of wars. Although it does not represent a guarantee of victory in the war, it has always interfered in the conduct of conflicts.[13] Furthermore, from the use of compound bows by Genghis Khan to the launch of nuclear bombs in World War II, technological surprises were disruptive as threats when associated with successful strategies. Consequently, the use of unmanned vehicles falls into the same category.[14]

From a conceptual point of view, the technological development of unmanned vehicles started "in 1898, [when] Nicola Tesla built a radio-controlled vessel, classified by him as a potential weapon." Extensive experimentation continued during the twentieth century, culminating with remotely piloted aircraft (RPA) capable of launching smart weapons.[15]

There are two explanations for the prevalence of military interest in unmanned systems. The first and most apparent is that the increase in their use worldwide occurred due to technological development, costs, and the persistence of conflicts. The second explanation, although subjective, refers to the fascination provoked by such systems to serve different interests of companies and citizens with different motivations. Engineers, philosophers, politicians, and especially military commanders have identified various advantages according to each application area.[16]

Parallel to the development of military RPA, there was also widespread use of commercial-grade RPA in various civilian activities, from border control to research, through surveillance and entertainment; concurrently, this has led to threats such as violation of privacy, interference with air traffic control, and security of installations, among others. Requirements related to these threats have become part of the risk analysis scenario due to the use of this equipment in public environments.[17]

Within this context, the conflict in Ukraine from 2014 until 2016 marked a milestone from the military point of view, with the introduction of many civilian RPAs adapted as weapons. The separatists and the Ukrainian armed forces used commercial equipment with different degrees of modification or adaptation, demonstrating the ability to use these RPA in both offensive and defensive operations.[18] These adaptations characterize the weaponization process.

The fundamental problem with this weaponization is that "the same weapons and ammunition purchased to be used against an external enemy can also be used against domestic opponents. However, the point we want to highlight is that the agents in charge of the government of the State are necessarily obliged to contemplate both scenarios simultaneously."[19]

Seeking to understand the effects obtained with the weaponization of SRPA, this study discusses the application of general systems theory (GST) to enhance or mitigate those effects in offensive or defensive military operations.[20] In addition, we provide data for improving the operational risk management process conducted in military planning, with the insertion of weaponized remotely piloted aircraft systems (sistema de aeronave remotamente pilotada, SARP) possible effects in the risk management process of Brazil's joint operations doctrine.[21] Thus, we present a theoretical approach for analyzing the use of weaponized SARP.

Development

After the First World War, the "passionate desire to avoid war determined the entire course and initial direction of the study" of international relations theory (IRT).[22] Analyzing the historical trajectory, van Creveld states that "following four and a half centuries of development that had started around 1300, the state found itself perhaps the most powerful political construct ever. Relying on its regular armed forces—first the military, then the police and the prison apparatus—it imposed order on society to the point that the only organizations still capable of challenging it were others of the same kind."[23]

Extrapolating this concept to the anarchic international system, Art and Waltz stated that the military powers would play a role as creators of order in chaos. Eventually, this order is achieved by the threat of force or its practical application, invariably with reflexes of international power balance. The study of the use of force in this scenario is linked to four central questions: "(1) What are the ways in which a state can use its military force?, (2) what determines how they use it?, (3) under what conditions are they likely to use it?, and (4) how can they control the competition in armaments that frequently arises among them?" When answering these questions, we can obtain a description of the strategic objectives of a given state, with a direct impact on military planning.[24]

Thus, with the respective evolution of IRTs, Waltz's work represented the theoretical framework of neorealism, seeking explanations for the conflicts of the post–Cold War scenario from a systemic view. "Any approach to international politics, so that it is properly called systemic, must, at least, try to infer some expectations about the behavior of the State and its interactions, from the knowledge of elements of systemic level."[25] For this reason, it is not possible to understand international politics just by looking at states, which risks limiting understanding of the scenario to simple descriptions. It would be necessary to constantly review the variables under study, given the complexity of the interactions, in an endless logical spiral. Any omissions in the study of international interactions would lead to a conceptual flaw that could compromise the result of the analysis.[26]

Waltz states that it is necessary to observe the interactions of the various actors, which can differ from the state structure, modifying the observed results. The number of variables and the number of possibilities of interactions exclude a Cartesian analysis based only on the state apparatus. In this way, the systemic view will show the logic and the global result of interactions and facts that could remain unexplained if they were analyzed in isolation. It is possible to understand why a conflict occurs, what to expect from it, how to resolve it, and, therefore, how to change the individual results.[27]

As Hartmann and Giles explained, the use of SRPA as a weaponized platform within reach of the population allowed for the potential of violence, given the characteristics of these assets.[28] In addition, van Creveld stated that "the rise of the state is inseparable from that of modern technology."[29] Therefore, conversely, studying the technical progress of military equipment will aid in understanding phenomena that can cause social impacts.

Being at the disposal of various actors—such as the general population, regular army, and irregular warfare forces—SRPA and their use confirm the definitions of Fourth Generation War, with the "return to a world of cultures, not merely states, in conflict."[30] The Fourth Generation War scenario increases uncertainty since "international politics can sometimes be considered a realm of accidents and convulsions, with sudden and unpredictable changes."[31] In this context, the operational risk management process seeks solutions to overcome such uncertainties, contributing to the success of military operations, whether offensive or defensive.

Therefore, this work studies the modeling of the effects that can be caused by weaponized SRPA and the proposition of including these SPRA in the operational tasks and activities of air and space control, as defined by the Brazilian air force basic doctrine.[32]

Theoretical Approach

We chose GST to study the effects obtained from the weaponized use of SRPA. Bertalanffy describes "three main aspects [to GST], which are not separable in content, but distinguishable in intention": systems science, systems technology, and systems philosophy.[33] The first refers to scientific research aimed at understanding scenarios with several variables that would cause an interdisciplinary confluence for understanding the elements of the scenarios and their interrelationships. In addition, the isomorphism of systems, even of different natures, can facilitate understanding processes. Systems technology refers to the development of mathematical or computational tools (or both) that would enable efficient solutions, according to an epistemology of its own and distinct, for example, from positivism. Finally, systems philosophy seeks the reformulation of references and perceptions, according to the concept of the "system" as a new scientific paradigm.

Those three axes of the GST theoretical framework enable a scientific approach regarding systems science due to the number of variables involved in understanding the use of weaponized SRPA in military planning. This "requires complicated mathematical techniques and computers to solve problems that far exceed the capacity of any individual mathematician."[34] Kauffman's NK Model[35]—initially created to study genetic interactions—was used because, as stated by Bertalanffy, "we often find formally identical or isomorphic laws in different fields."[36]

Finally, systems philosophy intends to improve airpower planners' situational awareness of the kind of threat that SARP represents. The point is to discuss the entire tactical balance framework, correctly understand the risk level, and contemplate using SRPA in all military planning scenarios. For example, in a system with a deployed air force base, the risk of a fighter jet being destroyed on the ground by a SARP must be considered in selecting the airdrome from the beginning of planning.

Thus, we used GST as a theoretical project for the weaponized use of SRPA, considering the difficulty of understanding, predicting, or controlling its global behavior, even with a reasonable amount of available data related to military planning enhancement. To reduce the complexity for the purposes of this study, the weaponized use of SRPA can be understood as a complex adaptive system of three variables: mass of explosives, type of explosives, and damage/lethality radius.[37]

Kauffman's NK Model was first used from a mathematical algorithm based on simulations and predictions of genetic interactions to study complex systems. Because the method simulates probabilities on complex data networks, it can be used in different systems of interest. As Gavetti and Levinthal explain, "the variable N refers to the number of distinct attributes in an overall policy choice. For instance, in a choice of a company business strategy, several decisions must be made, including decisions about how the product or service is to be marketed, such as issues of brand name and distribution channels, and how it is to be produced, such as the degree to which activities will be done within the firm or outsourced. The variable K refers to the extent to which the payoff associated with one policy choice depends on other policy choices."[38] Proving such flexibility, Frenken used the NK Model to support the study of complexity in innovation networks in the history of the aeronautical industry: "In the NK Model, the quality or fitness of an actor in the network is simulated using random values for the fitness of each actor, and the aggregated fitness of the network is calculated as the average f of the fitness values of all actors. The f^{XYZ} complexity of a network is indicated by K, which refers to the number of dependent relations within a network and has a minimum possible value of K and a maximum possible value of K = N-1."[39]

Frenken also pointed out the possibility of using the NK Model within a production engineering context to analyze the complexity represented by the interrelationships between different tasks necessary to produce a specific output.[40] In a correlated study, Assis proposed using Kauffman's NK Model to evaluate the result of interdependencies of variables within a business scenario or project for the management of multinational organizations, analyzing the interdependence among the head office and its respective subsidiaries.[41]

To meet the objectives of this work, the NK Model was adapted to the engagement scenario of weaponized SRPA. The interactions of a defined quantity of SRPA units will be simulated using the appropriate algorithm, with different employment profiles, from K=0 to K=N-1. Next, we carried out comparative simulations related to the employment profile of one A-29 Super Tucano aircraft as a control group.

It should be noted that, in the case of the employment of a single A-29, if "the value of K is low and there is little interaction among policy choices, then the fitness landscape is smooth or highly correlated. With a low value of K, a change in one policy has little impact on the fitness contribution of other choices."[42] For this reason, as Gavetti and Levinthal point out, when there are no interrelationships to be studied, "to ensure that the results reflect the underlying structure of the model and not merely particular realizations of a highly stochastic process, the results are based on the average behavior of organizations over 100 independent runs of the simulation model. For each of these runs, a distinct landscape is specified. Each of these landscapes has the same structure in terms of K and N but is seeded independently."[43]

The obtained data allowed us to carry out a comparative analysis between the use of a variable number of SRPA versus a single A-29. This comparison aims to verify the hypothesis "of equifinality, that is, that the same final state can be achieved starting from different initial conditions and in different ways."[44]

Standards

The present study adopted the standards described here to delimit a certain physical characteristic of the analyzed platforms. To obtain the SRPA parameters, we analyzed the following Brazilian government agencies linked to the standardization of the use of RPA: The FAB's Department of Air Domain Control (Departamento de Controle do Espaço Aéreo, DECEA) and the Brazilian National Civil Aviation Agency (Agência Nacional de Aviação Civil, ANAC).

DECEA is the central structure of the Brazilian Airspace Control System (Sistema de Controle do Espaço Aéreo Brasileiro, SISCEAB), which, in turn, is integrated with the Brazilian Air and Space Defense System (Sistema de Defesa Aeroespacial Brasileiro, SISDABRA). This peculiarity, therefore, optimizes the operational task of air and space control assigned to the FAB.[45] The DECEA states that "according to the International Civil Aviation Organization (ICAO), unmanned aircraft (UA) . . . are subdivided into three categories: Remotely Piloted Aircraft (RPA), Aero models and Autonomous. The first two have similar characteristics; they are unmanned and piloted from a remote pilot station. However, RPA, unlike aero model aircraft, will be used for non-recreational purposes . . . . Unmanned aircraft classified as autonomous have the characteristic of not allowing human intervention once the flight has started."[46]

Accordingly, "the RPA are classified according to RPA's maximum takeoff weight (MTW) by ANAC as follows: (1) Class 1: RPA with maximum takeoff weight greater than 150 kg (330 lbs); (2) Class 2: RPA with maximum takeoff weight greater than 25 kg (55 lbs) and less than or equal to 150 kg; and (3) Class 3: RPA with maximum takeoff weight less than or equal to 25 kg."[47] Therefore, for the scope of this study, SRPA is defined as a class 3 RPA platforms with MTW less than or equal to 25 kg.

For A-29 aircraft,[48] we used the following data to define the engagement profile: two-seat aircraft; three general purpose aerial bombs—500 lbs (MK-82 type); machine guns with 500 cartridges; altitude pressure at the base track of 2,000 ft; takeoff temperature 21º C (70º F) International Standard Atmosphere (ISA + 10º C); cruising altitude of 15,000 ft; the outside temperature at the cruise level -5º C (23º F); altitude pressure of 500 ft at the objective area; the temperature at target altitude 29º C (84º F) (ISA + 15º C); 5 minutes over the target; paved runway with 3,000 m (10,000 ft) length; and absence of obstacles for takeoff and landing. These conditions define, after rounding, a range of action of 150 NM, 120 minutes of flight time, and 500 kg (1,100 lbs) of fuel consumed.

The data produced could be incorporated into operational risk management in military planning, with the following qualifications, as stated in the Brazilian joint planning doctrine: "a) the identification of threats; b) the assessment of risks arising from hazards; c) the formulation of risk control measures; d) the assessment of residual risk; e) the risk decision; f) the implementation of 'risk control measures'; and g) supervision of the effectiveness of such measures."[49]

From the definition of those initial parameters, we made a preliminary analysis for the composition of this chapter.

Preliminary Analysis

According to Waltz's neorealistic view,[50] global stability would increase in bipolarity, tending toward a geopolitical balance. After the end of the Cold War—which caused the disruption of geopolitical bipolarity, as Lind discussed—states lost their monopoly of violence.[51] Thus, understanding the weaponization of SRPA as an enhancer of violence in Fourth Generation Wars demanded a theoretical framework capable of absorbing the large number of variables involved, leading to the selection of GST.[52]

Hypothetically, when Actor A uses SRPA against Actor B with the aim of causing an X effect, we can consider modeling such a scenario. Initially, within the context presented, Actors A and B may or may not represent state entities, demonstrating that a systemic analysis will be required. Subsequently, the magnitude of effect X will imply a wide range of possibilities: damage, neutralization, destruction, strategic paralysis, psychological effects, and so forth. In addition, the motivations inherent to the use of SRPA were not analyzed because they exceeded the scope of the present study, although they provide information on the expected level of performance: civilizational, territorial, economic, political, psychosocial, and so on. Finally, the conclusions or predictions obtained for analogous situations will guide the proposals of appropriate courses of action planned to enhance the effects of SRPA employment, offensive or defensive, typically used in military war games.[53]

Due to the existence of many variables, this hypothetical design led to a chain of interactions that, from a theoretical point of view, is equivalent to the use of a systemic analysis, such as GST,[54] which led to valuable data for risk management in military planning associated with using SRPA. In addition, we selected the NK Model. Its algorithms, used in computer simulations, provided the necessary data to outline the effects produced by the weaponized employment of SRPA.[55]

We corroborated the hypothesis of equifinality; that is, the same result can be achieved by different means, as in employing a variable number of SRPA or air-to-ground engagement of one A-29 aircraft.[56] The results showed that platforms built from the weaponization of SRPA should not be neglected in military planning scenarios, as they represent a significant threat to risk management.

Within this context, the data may support future studies on the imposition of order on chaos, answering the four questions raised by Art and Waltz on means, modes, circumstances, and control of the use of force (or the threat the use of force) resulting from the SRPA weaponization process.[57] These issues are invariably associated with operational risk management for military planning.

Conclusion

In attempting to understand the maintenance of peace,[58] several IRTs were created to explain the world scientifically and logically.[59] For proper contextualization, the present study demonstrated the nature of the historical progression of wars, going through the four generations of wars. Considering the Westphalia Treaty as the initial landmark of the states, the seventeenth and eighteenth centuries experienced the army's professionalization, characterizing First Generation War.[60] Later, from 1789 on, the Industrial Revolution consolidated the technological development observed in the First World War, which marked the advent of the Second Generation War.[61]

The increase of warfare mobility, first seen during World War II, shaped Third Generation Wars.[62] The Cold War ending, although showing Western dominance,[63] did not bring theoretically predicted peace, due to the multiplication of conflicts.[64] The consequent loss of the state's monopoly of violence configured the Fourth Generation War.

Considering that the development of the state cannot be dissociated from technological development,[65] the study of technical progressions, such as the weaponization of SRPA, is valued due to the optimization of the military planning process. Thus, the analysis of the weaponized use of SRPA presented several variables, which led to the selection of GST for constructing the theoretical approach.

In addition, with the NK Model, we proposed computer simulations for collecting data linked to the understanding and dimensioning of the effects obtained with the use of a variable number of SRPA, based on the comparison with the effects data from the engagement of one A-29 aircraft. The use of this aircraft, selected according to the concept of close-air support missions, was considered as a control group for the study.

So, the possibility of obtaining warfare effects within a Fourth Generation War scenario with actors that may be different from state entities can be enhanced by using weaponized SRPA.[66] This was proven by the corroboration of the hypothesis of equifinality, within which the same final state can be reached from different possibilities. At the same time, understanding this fact will contribute to efficient risk management analysis in military planning, starting from including SRPA as a threat in the scenarios to be studied.

Notes

1. Brazilian Ministry of Defense (MOD), DCA 1-1 DOUTRINA BÁSICA DA FORÇA AÉREA BRASILEIRA.
2. MOD, DCA 1-1.
3. Magnoli, *História das Guerras*, 164.
4. Lind, "Understanding Fourth Generation War," 12.
5. Magnoli, *História das Guerras*, 324–27.
6. Magnoli, 13.
7. Lind, "Understanding Fourth Generation War," 12.
8. Lind, 13.

9. Magnoli, *História das Guerras*, 386–93.
10. Schuurman, "Clausewitz and the 'New Wars' Scholars."
11. Miranda and Nascimento, *Conflitos Assimétricos*, 5
12. Lind, "Understanding Fourth Generation War," 13.
13. John, "Unmanned Systems in Perspective," 1–95.
14. John, 1–95.
15. John, 1–95.
16. John, 1–95.
17. Hartmann and Giles, "UAV Exploitation," 205-6.
18. Hartmann and Giles, 211.
19. Heye, "Democracia, controle civil e gastos militares," 111.
20. Bertalanffy, *Teoria geral dos sistemas*, 54–64.
21. MOD, *DOUTRINA DE OPERAÇÕES CONJUNTAS*.
22. Carr, *Vinte Anos de Crise*, 11.
23. Van Creveld, *The Rise and Decline of the State*, 183.
24. Art and Waltz, *The Use of Force*.
25. Waltz, *Theory of International Politics*, 50.
26. Waltz, *50*.
27. Waltz, 50.
28. Hartmann and Giles, "UAV Exploitation," 205–6.
29. Van Creveld, *The Rise and Decline of the State*, 377.
30. Lind, "Understanding Fourth Generation War," 13.
31. Waltz, *Theory of International Politics*, 65.
32. MOD, DCA 1-1.
33. Bertalanffy, *Teoria geral dos sistemas*, 15.
34. Bertalanffy, 22.
35. Kauffman, *The Origins of Order*.
36. Bertalanffy, 62.
37. Holland, *Hidden Order*.
38. Gavetti and Levinthal, "Looking Forward and Looking Backward," 118.
39. Frenken, "A Complexity Approach to Innovation Networks," 260.
40. Frenken, "Modelling the Organization of Innovative Activity," 12–16.
41. Assis, "INTERDEPENDÊNCIA ENTRE SUBSIDIÁRIAS ESTRANGEIRAS."
42. Gavetti and Levinthal, "Looking Forward and Looking Backward," 118.
43. Gavetti and Levinthal, 124.
44. Bertalanffy, *Teoria geral dos sistemas*, 112.
45. MOD, DCA 1-1.
46. MOD, "ICA 100-40 AERONAVES."
47. National Civil Aviation Agency, "Requisitos Gerais para Aeronaves," 5.
48. Embraer, *Manual de Voo*.
49. MOD, *DOUTRINA DE OPERAÇÕES CONJUNTAS*, 236.
50. Waltz, *Theory of International Politics*.
51. Lind, "Understanding Fourth Generation War," 13; and Miranda and Nascimento, "Asymmetrical Conflict and the State."
52. Bertalanffy, *Teoria geral dos sistemas*.
53. MOD, *DOUTRINA DE OPERAÇÕES CONJUNTAS*.
54. Bertalanffy, *Teoria geral dos sistemas*.
55. Kauffman, *The Origins of Order*.
56. Bertalanffy, *Teoria geral dos sistemas*.
57. Art and Waltz, The Use of Force.
58. Carr, *VINTE ANOS DE CRISE*.

59. Castro, *Teoria das relações internacionais.*
60. Lind, "Understanding Fourth Generation War," 13.
61. Lind, 13.
62. Lind, 13.
63. Schuurman, "Clausewitz and the 'New Wars' Scholars."
64. Miranda and Nascimento, "Asymmetrical Conflict and the State."
65. Van Creveld, *The Rise and Decline of the State.*
66. Lind, "Understanding Fourth Generation War," 13.

Chapter 2

Cyber Defense in the Brazilian Air Force
Overview and Future Perspectives

Pedro Arthur Linhares Lima
Gills Vilar Lopes
André Lucas Alcântara da Silva
Ines Correa Gomes Cardinot

Introduction

Cyberspace is a sensitive subject within the Brazilian Air Force (Força Aérea Brasileira, FAB),[1] a military organization in which technology plays an important role and that uses aircraft, space platforms, and systems that incorporate cutting-edge technology.[2] With the creation of the cyber strategic sector in 2008 by the *National Defense Strategy* (*Estratégia Nacional de Defesa, END*), Brazil joined the list of nations that consider cyberspace an environment to be protected, recognizing its mutable and volatile characteristics and the potential impact of this environment on several other sectors of national development.

This first step triggered a sequence of initiatives aimed at creating institutions, defining doctrines, and preparing and using cyber assets to provide the necessary security in the cyber environment. The head of the Institutional Security Office of Brazil (Gabinete de Segurança Institucional da Presidência da República) was responsible for the issue at the political level, addressing concepts such as information security and cybersecurity. The strategic, operational, and tactical levels fell under the direct responsibility of the ministry of defense and the armed forces.[3] In addition, *END* assigned the Brazilian army responsibility for developing cyber defense,[4] which led to the creation of the Cyber Defense Command (Comando de Defesa Cibernética, COMDCIBER), the central office of the Military Cyber Defense System (Sistema Militar de Defesa Cibernética, SMDC).[5]

This study was financed in part by the Coordination for the Improvement of Higher Education Personnel - Brazil (CAPES) – Finance Code 001.

However, that does not mean that protecting computing assets and securing communications is exclusively the Brazilian army's responsibility. On the contrary, both the Brazilian navy and the Brazilian air force, which are part of SMDC, are directly involved through their respective cyber defense offices.[6]

The strategic characteristics of cyberspace led the Brazilian air force to establish specific guidelines regarding the development of cyber defense capabilities and the use of cyberspace as an operational domain for aerospace power. The solution was the establishment of a cyber defense system (Sistema de Defesa Cibernética, SISDCAER) and the creation of its central office, the Brazilian Air Force Cyber Defense Center (Centro de Defesa Cibernética da Aeronáutica, CDCAER).

With this as background, the main objective of this chapter is to present an overview of how the FAB's cyber defense is structured, especially from a strategic point of view. A survey reviews open sources to illustrate the importance of the cyber environment for FAB missions.

This chapter is divided into three parts. The first describes a normative view about cyber defense, paving the way for strategic and prospective thinking in this environment. The second part deals specifically with CDCAER, and the last points to some future perspectives and the intersection between cyber and space power in the context of a multidomain environment.

Normative Aspects of Cyber Defense in Brazil

The *END* outlines the national defense dimensions of the cyber sector, also including cyber as basic means in all armed forces and as a basic element to enable national and military development in Brazil.[7] Regarding the availability, integrity, confidentiality, and accessibility of critical information infrastructures,[8] the normative aspects of cyber defense go hand in hand with information security,[9] determining the guidelines used to protect against vulnerabilities as well as the best practices to follow and their applicability regarding the singularity of each strategic sector—cyber, nuclear, and space.

Decree no. 5484/2005 approved the first version of the *National Defense Policy (Política Nacional de Defesa, PND),*[10] which was followed by the *END,* first published in 2008 and later updated in 2012. Both documents constitute the first normative aspects of national defense that give greater coverage at the political and strategic level and are aimed at the Brazilian armed forces.

According to the *END*, cyber capabilities were to be employed for industrial, educational, and military uses. The communication technologies of all contingents of the armed forces were included as a priority to ensure their capacity to act in networks and benefit communication between the armed forces and space vehicles. In the cyber sector, an organization was to be created to develop cyber capabilities in the industrial and military fields.

In its 2018 version, *END* addresses the intersection of the armed forces and cyber defense. This document also mentions the influence of the cyber sector in new expressions of power, such as hybrid war, and its central role as a dissuasive power.[11]

In accordance with the 2018 version of the *PND*,[12] the cyber environment requires special attention in relation to security and defense in this virtual space, composed of computational devices connected (or not) in networks in which digital information—essential to ensuring the functioning of information management, communications, and command and control systems (C2)—is exchanged. A significant portion of military activities, in particular air and space forces, depends on these systems.

The Brazilian Air Force Cyber Defense Center

CDCAER comprises two main areas of focus. The first is related to FAB's central office of defense and cyber war activities. The second deals with CDCAER's effective presence in SMDC. CDCAER is responsible for planning, coordinating, and centralizing protection, exploitation, and cyberattack actions. In addition, it has the highest level of cyber situational awareness within the FAB. From a strategic point of view, CDCAER is responsible for planning and managing cyber defense training and actions. Other strategic responsibilities of CDCAER include coordinating FAB's critical infrastructures and digital systems security, the study of new cryptographic technologies, and establishing lines of research for the cyberspace field.

From the operational perspective, CDCAER responsibilities center on the orientation and execution of cyberspace protection, exploration, and attack and on the preparation and use of aerospace military power. In addition, the CDCAER operates the Network Incident

Response Center (Centro de Tratamento de Incidentes de Rede, CTIR-FAB) and supervises the Network Incident Response Teams (Equipe de Tratamento e Resposta a Incidentes em Redes Computacionais, ETIR), distributed in other air force branches. Finally, from the administrative point of view, CDCAER is responsible for establishing and supervising SISDCAER governance, inspecting system compliance, and coordinating other systems in the FAB.[13]

Additionally, as a branch of SMDC, CDCAER represents the Brazilian air force in issues related to cyber defense and acts at the strategic level. It collaborates with other SMDC activities through coordination with the central office—in this case, COMDICIBER—and with cyber intelligence activities for the benefit of SMDC and the defense intelligence system.[14]

The next section points out some contexts and future threats that may represent great challenges for CDCAER duties.

A Look to the Future: Multidomain Dimensions and Intersection with Space

Battling different types of threats to Brazilian aerospace power, including cyberattacks and cyber espionage, requires systematic investments in science and technology. In this context, the multidomain dimension—which involves the integration of the aerial, space, and informational domains[15]—can be seen in some products and projects the FAB recently developed or acquired.

For example, Embraer KC-390, the largest military aircraft built in the Southern Hemisphere and intended for military transport and in-flight refueling, has advanced avionics and electronic warfare systems, such as self-protection system, heads-up display, and directional infrared countermeasures.[16] Therefore, this aircraft constitutes a true air force vector that aggregates high technology and electronic and cyber capabilities. In view of such versatility and technological efficiency, it may become a target for not only industrial espionage but also cyber espionage.

Six years ago, the FAB signed a contract with Saab, a Swedish company, to purchase 36 Gripen-NG fighters and the corresponding technology transfer to produce these new generation aircraft independently and within Brazil. The so-called FX-2 Project is an ambitious program to modernize Brazil's fighter fleet. One of the attention-getting technologies of these fighters is the advanced artificial intelligence (AI) that works autonomously in several areas simultaneously and provides suggestions to the pilot. The AI's suggestions can vary from weapon selection to maneuvers; further, the AI's "avionics intelligent architecture" allows old algorithms to be replaced by new ones without reducing the complete availability of the aircraft.[17] The fact that the Gripen collects, stores, and transmits sensitive data—and therefore needs to have its software updated—reminds us of the famous case of the Stuxnet worm in Iran's supervisory control and data acquisition nuclear systems. It demonstrates how cyber intelligence and the development of cyber weapons can compromise infrastructure and critical assets, even when not connected to external networks.

Another example took place in 2017, when the first 100 percent Brazilian-controlled geostationary satellite was launched to meet telecommunication demands related to public policies, strategic security, and national defense needs. It is, therefore, an important spatial asset to guarantee surveillance of Brazil's vast national territory and to promote strategic communications in places of difficult access, such as the Amazon region, through the Ka band. The X band used in the satellite is aimed at encrypted military and government communications in addition to transferring georeferenced images. As the satellite is dual-use, civilian personnel from Telebras (Brazilian telecom) and Brazilian military personnel share the space operations center headquarters in the federal capital.

Because these new space assets (satellites, space stations, constellations, etc.) need cyberspace to communicate with each other and with Earth, gaps open for cyber threats (such as worms, viruses, man-in-the-middle, social engineering, etc.). To get an idea about this cyber-and-space issue, the new US Space Force focuses on how the satellites can be used against themselves (by software) and against other nations' satellites (by hardware).

The weaponization of outer space has been carried out especially by the world powers. For example, in January 2019, the US Air Force (USAF) National Air and Space Intelligence Center released an overview with very realistic and pessimistic views on the strategic use of military capabilities in space. Also in this report, the USAF emphasized cyber threats arising from the development of offensive capabilities that can enable the enemy to take military advantage from space. This would be possible, for example, using cyberattacks—for instance, worms like Stuxnet—and techniques—like social engineering—aimed at assets that could be in any environment, such as space, earth, and air.[18]

Managing cyber protection of space assets in Brazil—such as space platforms, personnel, aircraft, weapons, installations, equipment, and systems—requires not only investment in cutting-edge science and technology but also evaluation of public policies related to cyber defense.

Conclusion

To fulfill its constitutional mission, the FAB uses air vectors, equipment, and cutting-edge technologies. Such sensitive technologies represent a major concern, as they can compromise national defense if they are not well managed and monitored.

Cyber means are more and more central to achieving state department objectives. Thus the use of cyber means has become a new dimension to be considered in conflicts. Aware of this new reality, Brazil started to organize and structure to face this new challenge with the publication of END in 2008. Likewise, the Brazilian air force started planning and developing cyber defense capabilities to protect its critical assets and operate in this new domain. The creation of the CDCAER was proposed to become SISDCAER's central office and take charge of planning, coordinating, and centralizing cyber activity. From the operational point of view, CDCAER will be responsible for guiding and executing cyber protection, exploitation, and attack; operating the Network Incident Response Center; and supervising ETIRs distributed across the military organizations.

Looking toward the future, CDCAER must be very well organized, prepared, equipped, and able to use cyberspace and operate in the multidomain dimension to protect new aerial and space platforms against cyberattack, because otherwise, Brazilian aerospace sovereignty may be seriously compromised.

Notes

1. Ministry of Defense (MOD) and Air Force Command, "Concepção Estratégica."
2. Brazilian National Congress, *National Defense Policy and National Defense Strategy, 2012.*
3. MOD and Joint Chiefs of Staff, *Doutrina militar de defesa cibernética.*
4. Brazilian National Congress, *National Defense Policy* and *National Defense Strategy, 2012.*

5. MOD, DCA 1-1.
6. Ferreira, "Mudanças Na Política de Defesa Cibernética Brasileira"; and MOD, DCA 1-1.
7. Except for the Brazilian Army, where it is also designated as the end area.
8. Mandarino, *Seguranca e Defesa do Espaço Cibernético Brasileiro*, 39. According to the author, critical information infrastructures are extremely necessary for the continuity of critical infrastructure services in a country. Examples include the information and telecommunications sector, computers, satellites, and fiber optics.
9. Mandarino, *Seguranca e Defesa do Espaço Cibernético Brasileiro*.
10. Chamber of Deputies, Decree 5484.
11. Brazilian National Congress, *National Defense Policy* and *National Defense Strategy (Draft), 2016*.
12. Brazilian National Congress.
13. MOD, DCA 1-1.
14. MOD and Air Force Command, Decree 1008/GC3.
15. Lonsdale, "Information Power," 137–57.
16. AEL Sistemas, "Sistemas Para o KC-390."
17. Saab, "Gripen E."
18. National Air and Space Intelligence Center, "Competing in Space."

Chapter 3

SIPAM and SIVAM Projects
A Challenge in the Amazon

Luiz Fernando Póvoas da Silva
Flavio N. H. Jasper

Introduction

This chapter analyzes the Amazon Protection System (Sistema Integrado de Proteção da Amazônia, SIPAM) and the Amazon Surveillance System (Sistema de Vigilância da Amazônia, SIVAM). In the nearly two decades since their conception and implementation, these systems maintained Brazil's sovereignty in its territory and airspace and in its strategic surroundings, the Amazon Arc. These challenges will be examined from two perspectives: territorial sovereignty and airspace sovereignty.

The Amazon region is admittedly one of the richest areas in the world. As a source of sufficient natural, mineral, and water resources, the Amazon has the capacity to make Brazil a world economic powerhouse; at the same time, it is the geopolitical base for the financial, industrial, agricultural, commercial, and technological contentiousness generated by Brazil's emergence among developed countries. The riches of the Amazon—the water, the mineral resources, the biodiversity, the natural products, and the people—must stop being just listed as resources to enter the productive chain of the region and the country.[1]

Freire highlights the problems Brazil faces due to the riches of the Amazon region, emphasizing that as a country of continental dimensions, Brazil also has proportional problems, mainly in the Amazon, which has become the target of the greed of developed nations due to the grandeur of its natural resources and minerals.[2] The Amazon region, with its mineral and natural wealth, can contribute a great deal to the Brazilian national development project. However, Freire points out that it is important to present the differences between growth and development. "Growth has quantitative connotations, while development has qualitative connotations," which are represented by wealth, regional transformations, the evolution of society, and environmental preservation.[3]

The factors of growth versus development were the focus of the conference held in Rio de Janeiro in 1992, which covered environmental problems, illegal mining, and indigenous peoples. The conference resulted in a concept of sustainable development that made it necessary to invest in the environmental management of countries with actions aimed at protecting natural resources, establishing protection areas, and combating deforestation and desertification through investment in technologies capable of reinforcing the environmental management of countries.[4]

These were the factors that led the Brazilian government to implement SIPAM, highlighted in Ministerial Directive No. 003/2002, of March 4, 2002, which aimed to integrate, evaluate, and disseminate information for planning and coordination of global government actions operating in the Amazon, aiming to contribute to the sustainable development of the region.[5]

In Lia Machado's view, the current world, despite being more dynamic, is still full of uncertainties, one of which is the issue of the territorial state. This issue has three components, beginning with the institutional, that is, the form of control of limits and borders within the scope of institutions. The second would be conjunctural, "the sliding of frontiers into the national territory," characterized by the multiplication of special territories, a reference by the author to the demarcation of indigenous territories; and the third would be the structural aspect, that is, the use of a legality/illegality statute and the role of political boundaries in a world economy.[6]

For Almeida, the issue of sovereignty over the Amazon is a highly sensitive issue for Brazil. The reorientation of the central axis of defense issues in Brazil in relation to the Amazon demonstrates its importance for national policy, leaving aside old concerns with the southern border to shift the focus to the security of the Brazilian north.[7] Santos points out that SIPAM and SIVAM were born from the need for the federal government to adopt strategic actions that would allow the generation of up-to-date knowledge about the Brazilian Amazon, in addition to systematizing control, inspection, monitoring, surveillance, and protection of the region.[8]

Emerging from the technical-scientific-informational environment, SIVAM foresaw the implementation of a surveillance system based on network technology, satellites, and radars. Santos points out that SIVAM was a project designed to strengthen the argument that international demands for the preservation of the environment and the demarcation of indigenous lands represented a threat to national sovereignty, having as a fundamental reason the pressure from Western powers and nongovernmental organizations regarding the possible devastation of the Amazon rainforest.[9]

SIVAM has a symbolic dimension besides its commercial value, emphasizing technological modernity and the need for a more complex territorial control, without adopting the assumptions of the classic geopolitical models of territorial perimeter defense through occupation. Lia Machado explains that SIVAM was based on a conception different from the previous proposal, the *Calha Norte* Project, almost a classic militarization model, which had been suspended due to accusations, external and internal, of militarization of the border.[10]

To understand all these interrelationships and how they are connected to SIPAM and SIVAM, the concept of both projects will be discussed below.

The Conception of SIPAM and SIVAM Projects

The Integrated Amazon Protection System—SIPAM

SIPAM was born from the need to make the state's presence effective in the Legal Amazon by a system that would contribute to the control, inspection, and monitoring of this strategic region for Brazil.[11] The system was the result of the Explanatory Statement by the Ministry of Justice, along with the Ministry of Aeronautics (Ministério da Aeronáutica, MAER) and the Secretariat of Strategic Affairs (Secretaria de Assuntos Estratégicos, SAE), No. 194, of September 21, 1990.[12] Fernando Henrique Cardoso, Brazil's president, included, among others, the following proposals:

a) To authorize the SSA to formulate and implement a national coordination system, aiming at the integrated action of governmental agencies in the repression of illicit acts and environmental protection in the Amazon.

b) Authorize the Ministry of Aeronautics to implement the Amazon Surveillance System, integrated to the National Coordination System to be formulated by the SSA.

c) From the Explanatory Memorandum, the initial name was Amazon Surveillance System (SIVAM), an air force system, but the document also highlights that this system would be integrated to a national coordination system yet to be designed by the SAE. This system would be SIPAM.[13]

The Deliberative Council of the Amazon Protection System was created in 1999 with a purpose of establishing guidelines for the coordination and implementation of government actions, in accordance with the integrated national policy for the Legal Amazon.[14] In 2002, by means of Decree 4200, of April 17, the Executive-Secretary of the Deliberative Council of the Amazon Protection System was renamed Operational and Management Center of the Amazon Protection System (Centro Gestgor e Operacional do Sistema de Proteção da Amazônia, CENSIPAM), whose purpose was to promote the protection and sustainable development of the Legal Amazon, based on the integration of information and generation of knowledge about the Amazon region.[15]

CENSIPAM, besides its technological infrastructure, consists of three regional centers based in Manaus, Belem, and Porto Velho and has established partnerships with the Brazilian Institute of Environment (Instituto Brasileiro do Meio Ambiente e dos Recursos Naturais Renováveis, IBAMA); federal police; Chico Mendes Institute for Biodiversity Conservation the Brazilian government environmental agency, created by law 11.516, of August 28, 2007; National Indian Foundation (Fundação Nacional do Índio, FUNAI); civil defense; armed forces; and state and municipal governments.[16]

The participation of the armed forces was dictated by a normative document, Directive of the Ministry of Defense No. 003/2002, of March 4, 2002,[17] in which the armed forces mission was established: "To participate in the process of SIPAM activation jointly with the other instances of the public administration, in order to contribute to the operationalization of this system, in accordance with the National Integrated Policy for the Legal Amazon." One of the important aspects established in the directive was that the data entry of essential information and the execution of training for activation of the system was the responsibility of the Commission for the Coordination of the Amazon Surveillance System Project (Comissão Coordenadora do Projeto Sistema de Vigilância da Amazônia, CCSIVAM) of the Aeronautics Command.[18]

SIVAM preceded SIPAM, since it was already part of the Aeronautics planning for monitoring the national territory. However, the importance of the Amazon Region imposed itself as of 1990, causing the Brazilian government to set in motion actions that would allow for monitoring and control of this vast region.

The Amazon Surveillance System—SIVAM

In the 1990s, together with SAE and the Ministry of Justice, the MAER presented Explanatory Memorandum no. 194 prioritizing the problems in the Legal Amazon. The Explanatory Memorandum was approved by President Fernando Henrique Cardoso on September 21, 1990, with the following determination: the Ministry of Aeronautics is responsible for implementing SIVAM and integrating it into the national coordination system.[19]

Article 3 of the Decree of October 18, 1999, defines SIVAM as part of SIPAM and highlights that this system "aims at the execution of works and services, the acquisition of equipment and the allocation of goods destined to the collection, processing, production and dissemination of data on the Amazon, within the scope of SIPAM."[20] Therefore, the SIVAM project was born under MAER coordination and had, among others, the following responsibilities: generate updated knowledge on the Brazilian Amazon, create conditions for the integration of the government sector bodies in search of solutions to protect the Amazon, and systematize the control, inspection, monitoring, and surveillance of the region.[21]

To enable the implementation of SIVAM, MAER assigned the Commission for Implementation of the Airspace Control System (Comissão de Implantação do Sistema de Controle do Espaço Aéreo, CISCEA) the task of elaborating the basic and execution projects for the implementation of SIVAM. This was because, in the governmental structure, CISCEA was the only agency with competency and experience for the installation, in many areas, of cutting-edge technologies with multidisciplinary nuances. To conduct the project, CCSIVAM was created, coupled to CISCEA, and conducted by the same group of people.

Thus, SIVAM was born to be responsible for the collection, processing, and distribution of data to users who embraced SIPAM, providing conditions for a new style of integrated administration in the

Amazon region, with the support of equipment, software, and personnel for the collection of data in that area.[22]

Santos points out that SIVAM is made up of a complex network of instruments and technological resources, with information being its basic element and also a data acquisition and information processing system for government agencies in the Amazon region constituting a sophisticated integrated telecommunications and control network covering the entire air and land space of the Amazon.[23] The concept of the SIVAM project can be seen in figure 3.1; its initial purpose was to enable the technical implementation of SIPAM and only after its conclusion would it become CINDACTA IV. The keyword of both projects was integration, represented by radars, telecommunications stations, aircraft, satellites, and meteorological stations, as depicted in figure 3.1.

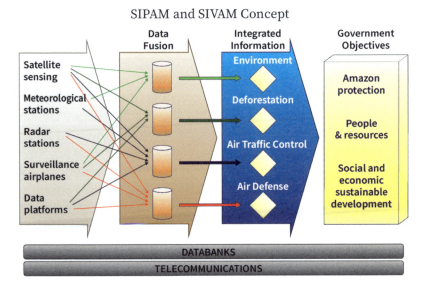

Fig. 3.1. SIPAM and SIVAM concept. (Source: the authors)

To finance this project, Brazil, with the participation of Banco do Brazil, took out a loan of US $1,022,800 from EximBank. The total cost of implementing SIVAM was US $1.395 billion, of which US $1.285 billion was allocated to equipment and services and US $110 million to civil works. Of the total investments, Raytheon, winner of the contract for the supply of equipment, received US $239.2 million.[24]

The contract between CCSIVAM and Raytheon, ATECH Foundation, and Embraer came into effect on July 25, 1997.[25] The executive arm of CISCEA for the implementation of SIVAM was the ATECH Foundation agency in charge of the receipt of goods and services as well as the integration of all equipment. Embraer was the national company chosen to supply the aerial surveillance and remote sensing aircraft.

With the inclusion of the air surveillance and early warning aircraft R 99 (AEWC), Brazil became part of the limited group of countries that had a functioning command-and-control air defense system.

SISDACTA: The Origin of SIVAM

SIVAM was the continuation of the project conceived as part of airspace control in the 1970s, the Integrated System of Air Defense and Air Traffic Control (Sistema de Defesa Aérea e Controle de Tráfego Aéreo, SISDACTA), which was the result of a working group established by the Aeronautics general staff through Ordinance No. R004, February 6, 1969.[26] The working group created SISDACTA in the 1970s in response to the need to integrate means and agencies that belonged to the other armed forces, such as the navy and army, as well as the need to manage and rule civil aviation traffic. Another alternative presented was that SISDACTA would be more economically viable whether the air traffic management structures and air defense were integrated (see fig. 3.3).[27] Therefore, one of the main goals presented by the working group was to integrate resources, that is, the operational infrastructure of visualization (radars), communication, software (air defense and air traffic control), besides the human resources to support and operate the system that was to be managed by a single agency, the Aeronautics Command.[28]

To meet the criterion of economic viability, one of the important aspects of the project was the plan to implement the system by modules (in stages),[29] a criterion that established the balance between the resources necessary for a project of this magnitude compatible with the country's financial capacity (see fig. 3.2).[30] Reflecting the concept of the working group, Pereira points out that the effective and gradual implementation of SISDACTA represented a great advance in national air defense, having enabled the Brazilian Air Force (Força Aérea Brasileira, FAB) to carry out two important functions for the country: the control of Brazilian airspace and air defense in a radar, telecommunication, and computer-only network that enabled the desired integration between equipment, functions and human resources, in addition to a considerable cost reduction.[31]

The country was divided into four air defense and traffic control regions as figure 3.2 shows. This division was also a result of the studies of the working group that presented five solutions, but all of them proposed to divide the country into air defense regions, coinciding with the existing flying information regions in the country.

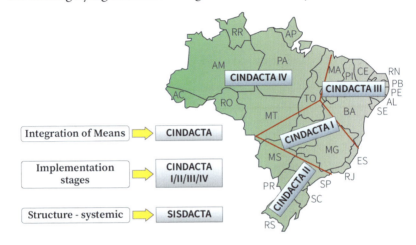

Fig. 3.2. SISDACTA's Concept. (Source: the authors)

The implementation of the project by stages started with CINDACTA I and finished with CINDACTA IV. As executive bodies of this system, their areas of responsibility are also shown in figure 3.2. Integrating these systems was necessary not only by the infrastructure requirements (radars, telecommunications, etc.), but also by the operational factor, that is, the integration between the military part of the system (air defense) and the civilian part (air traffic management [ATM]), making it possible to achieve excellence in flight safety.

This concept revolutionized the thinking of the time because all over the world, air traffic control and aerospace defense systems were separated. Today, this Brazilian concept is the global standard.

CINDACTA: First Step Toward Integrated Systems

For SISDACTA to be operational and function as designed, it was necessary to assemble a physical structure to meet the defined criteria, called the Integrated Center for Air Defense and Air Traffic Control, (Centro Integrado de Defesa Aérea e Controle de tráfego Aéreo, CINDACTA), which was regulated by Decree No. 73.160, of November 14, 1973. Article 1 established that CINDACTA's mission was to monitor and control air circulation as well as to conduct aircraft whose mission was to maintain the integrity and sovereignty of Brazilian airspace in the area defined as its responsibility.[32] Thus, CINDACTA had specified, in its regulation, nine competencies that, in short, aimed at managing, controlling, and coordinating the responsibilities involving air traffic and air defense.[33]

Later, SISDACTA evolved and was divided into two systems. The Airspace Control System (Sistema de Controle do Espaço Aéreo Brasileiro, SISCEAB) falls under the Department of Aerospace Control and is responsible for equipment maintenance and operation and for the ATM operational function. The Brazilian Aerospace Defense System (Sistema de Defesa Aeroespacial Brasileiro, SISDABRA), overseen by Aerospace Operations Command (Comando de Operações Aeroespaciais, COMAE), is responsible for military and aerospace defense operations and the airspace policing mission, with its specific software. The split of the systems is depicted in figure 3.3.

Air Defense System and Air Traffic Control

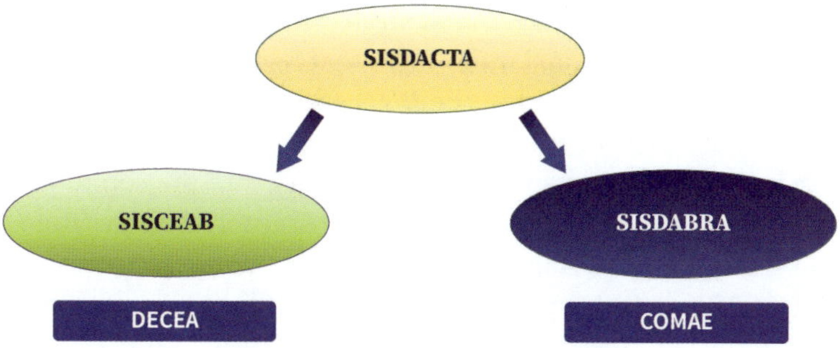

Fig. 3.3. Evolution of SISDACTA. (Source: the authors)

Although the system has evolved and divided into two specific systems, SISDABRA with aerospace defense and SISCEAB with air traffic management, both systems remain under the responsibility of a single agency, since COMAE (aerospace defense) and DECEA (air traffic control) are bodies belonging to the COMAER's organizational structure.[34]

Challenges and Threats: Present Impositions

In 2021, the Brazilian Minister of Defense expressed his concerns about the Amazon: "The current geopolitical scenario presents new challenges, such as environmental issues, cyber threats, food security, and pandemics."[35] The *National Defense Policy* (*Política Nacional de Defesa*, PND) also highlights this concern by defining Brazil's priority interests as being its strategic environment, which includes South America, the South Atlantic, the countries of the West African coast, and Antarctica. From the point of view of defense, in addition to the regions where political and economic powers are concentrated, priority must be given to the border strip, the Amazon, and the South Atlantic.[36]

The *National Defense Strategy* (*Estratégia Nacional de Defesa*, END) highlights that the Amazon is an area of geostrategic interest for Brazil. The protection of biodiversity and mineral and water resources, in addition to the energy potential in the Brazilian territory, is a priority for the country. The exploitation and socioeconomic development of the Amazon—in a sustainable way—will continue to be vital for national integration, requiring the increase of capacities to provide security and sovereignty, intensifying the military presence and the effective action of the state, preventing exogenous entities from influencing local communities. In this regard, the END clarifies that dissuasion should be the first strategic posture to be considered for the defense of national interests in that region.[37]

The concerns expressed in the PND and END are in line with the criteria established in the 1990s for the creation of SIPAM and SIVAM: concern for the environment, security, and the country's sovereignty over its territory, especially the Amazon Region. The PND emphasizes the importance of borders that require attention since people and goods transit through them, integrating regions and bringing the country closer to its neighbors. At the same time, illicit activities (organized crime, large-scale transnational drug trafficking, and violent civil war between criminals and the police) assume a transnational nature, requiring constant surveillance, coordinated action between defense and public security agencies, and close cooperation with neighboring countries.[38] The current conflicts have been characterized by a significant dissonance between the expectations of public opinion, the intention at the political level, and the perception of military leaders.[39]

The *PND* also stresses that, along with the continental and maritime dimensions, the aerospace environment is fundamentally important for national defense. The use of outer space, the control of Brazilian airspace, and its permanent articulation with that of neighboring countries as well as the continuous development of aerospace activity are essential to safeguard sovereignty and national interests.[40]

In this context expressed by the Minister of Defense and in the *PND*, Brazil needs to control its territory and borders, demonstrating the importance of the country having a system that allows it to exercise this control in terms of sovereignty over its territory, borders, and airspace as well as aspects of environmental control that include the problems of deforestation and illegal mining, among others—fundamental criteria for the creation of SIPAM and SIVAM. Illegal mining, for example, is not only a problem in Brazil but also affects a European country, France, which is obliged to use its armed forces to try to control illegal mining in its overseas possession, French Guiana.[41] The French legislature even created a commission of inquiry to address the matter, whose suggestions were presented through Report No. 4404 published in 2021.[42]

Regarding surveillance of Brazilian borders in the Amazon, for example, control over drug trafficking is currently carried out by COMAE through CINDACTA IV (which, as previously explained, was SIVAM). This control is carried out daily and in coordination with federal police to not only prevent the flight of aircraft transporting drugs but also to lead to the capture of traffickers. News published on the COMAER portal highlights that, on February 19, 2022, an aircraft had been detected in the central region of Amazonas without a flight plan and was on a known route for drug trafficking, being then subject to airspace policing measures of the Brazilian government. The news also highlighted that the aircraft made a forced landing in a sugarcane field near the city of Presidente Figueiredo; an H-60 Black Hawk helicopter from the FAB landed at the site with a federal police team that seized the illicit drug load. The report also points out that the commander of COMAE commented on the success of the joint operation, highlighting that the action proved not only the operational readiness of the FAB but also the capabilities of SISDABRA that allow consistent surveillance 24 hours a day, seven days a week.

This action was part of Operation Ostium to curb illicit acts in Brazilian airspace, in which the FAB and a public security agency, in this case the federal police, worked together.[43] Operation Ostium carried out 652 interceptions of aircraft in the last three years. This surveillance resulted, in 2021, in the seizure of 1,680 kg (3,700 lbs.) of drugs, and, in 2022, the action led to the seizure of more than a ton. Since 2017, the total amounts to 7.2 tons of seized drugs.

Likewise, the FAB acts, through Operation Yanomami, in the control of illegal mining in indigenous lands, the core of the SIPAM and SIVAM systems when they were created.[44] Illegal mining, deforestation, and drug trafficking are instruments that can be used as weapons to destabilize nations politically, economically, and militarily, particularly in the strategic Amazonian environment, which—with its instabilities, actors, and threats—is characterized as a multidomain scenario subject to offensive actions of all types as existing in hybrid warfare. Hybrid war can be conceptualized as asymmetric conflicts provoked externally for purposes based on economic interests capable of interrupting transnational projects through ethnic, religious, regional, political, geographic, and socioeconomic conflicts.[45]

As noted previously, the *PND* and *END* define the national defense strategy for the Amazon strategic environment as the strategy of dissuasion and for the employment of military power. According to Brazilian aerospace power scholar Lieutenant Brigadier Murillo Santos, dissuasion strategy requires attack capability, even if it is a ready-response strategy, otherwise it becomes a resistance-defensive strategy.[46]

Dissuasion as a national defense strategy transits in this current war environment, called multidomain due to the plurality of capabilities that compose the present relations of force and, also, due to the wide distribution of power in the social environment. The modern-day battlefield will be multidimensional. It incorporates the infosphere, enabling anyone—media, governments, individuals—to harness telecommunications technologies and cyber resources to extend their reach.[47]

In South America, modern conflict also bears traces or characteristics of the region's underdevelopment, such as low levels of educational and economic attainment, in addition to assuming the peculiarities of the continent as a producer of cocaine and marijuana. This is pointed out by the *PND* when the document highlights that, from a defense viewpoint, the possibility of armed conflicts in South America cannot be disregarded, so that Brazil may be motivated to contribute to the solution of eventual regional disputes or even to defend its interests.[48]

The existing challenges and threats in the strategic Amazonian surroundings are issues that should be faced by Brazilian society with its civilian agencies and military bodies; civil-military cooperation, through public policies; and interagency operations with the aim of maintaining the integrity of the national territory against any kind of transnational illicit activities. The challenges present are the almost the same as those of the 1970s, when SISDACTA was conceived, and later when SIPAM and SIVAM where conceived and implemented to deal with the challenges of the Amazon Region. These systems have made a great contribution allowing the country to apply its sovereignty over its entire territory and air space contributing to the Brazilian strategy of dissuasion.

Conclusion

This research traced SIPAM and SIVAM systems and how they counter the challenges in Brazil's strategic surroundings. It also demonstrated the enormous current challenges and threats in the strategic Amazonian surroundings, a large area in a multidomain environment that today could be considered a hybrid war scene.

First, the research highlighted why SIPAM and SIVAM were created, having as basic principles surveillance and control over the territory and over Brazilian airspace. Among the highlighted problems that were established as criteria for surveillance and control were deforestation, illegal mining, mainly in indigenous lands, and drug trafficking in the Amazon region.

The chapter also demonstrated that the concept of integrating material and human resources present in the creation of SISDACTA was also one of the basic criteria for the creation, in the 1990s, of the SIPAM and SIVAM projects. In these two projects, the concept of integration expands to include the integration of information and operations with other bodies such as the federal police, IBAMA, and FUNAI, among other agencies.

Bringing the analysis to the present day, the chapter highlighted that both the *PND* and the *END* also expressed concern with the Amazon Region, corroborating the need for the existence of the SIPAM and SIVAM systems, today called CINDACTA IV, part of SISDABRA, the system responsible for Brazil's aerospace defense.

In this update, the survey also showed that interagency cooperation remains, especially with the federal police in terms of airspace surveillance for the control of aircraft transporting drugs in the Amazon region, in addition to FAB cooperation in combating illegal mining activities on indigenous lands through Operation Yanomami.

The research also sought to highlight the importance of creating SIPAM and SIVAM to contribute to the dissuasion strategy advocated by the *PND* and *END* within the principles of protecting national territory, mainly the Amazon, as well as integrating public policies in the sense of enabling the sustainable development of this important region of Brazil.

This research is expected to contribute to other academic works that deal with these important and exciting themes that involve the Amazon region.

Notes

1. Freire, "REGIÃO AMAZÔNICA."
2. Freire, 2.
3. Freire, 5.
4. Ignacio, "ECO-92."
5. Ministry of Defense (MOD), "Ministerial Directive no. 003/2002."
6. Machado, "Limites e Fronteiras," 7–8.
7. Almeida, "Sistema de vigilância da Amazônia," 44.
8. Santos, "Strategic Information Management as a Conditioning Factor," 95.
9. Santos, 95.
10. Machado, "Limites e Fronteiras," 7.
11. MOD, "Revisão Do Planejamento Estratégico Da CENSIPAM."
12. Ministry of Justice, Ministry of Aeronautics, and Secretariat of Strategic Affairs, "Exposição de Motivos N°194," 18327.
13. Ministry of Justice, Ministry of Aeronautics, and Secretariat of Strategic Affairs.
14. MOD, "Revisão Do Planejamento Estratégico Da CENSIPAM," 6.
15. Civil Office, Decree 4200.
16. MOD, "Censipam Institutional Folder," 7.
17. MOD, "Ministerial Directive no. 003/2002."
18. With the creation of the Ministry of Defense in 1999, the Ministry of Aeronautics became the Aeronautics Command, subordinated to the Minister of Defense.
19. The Coordination System to be created would be the Integrated Amazon Protection System (SIPAM).
20. Civil Office, Decree of October 18, 1999.
21. Coordinating Committee for the Implementation of SIVAM Project, "O Sistema de Vigilância Da Amazônia (SIVAM)," 2–5.
22. Coordinating Committee for the Implementation of SIVAM Project, 2.
23. Santos, "Strategic Information Management as a Conditioning Factor," 119.
24. Almeida, "Sistema de vigilância da Amazônia," 45–47.
25. Coordinating Committee for the Implementation of SIVAM Project, "O Sistema de Vigilância Da Amazônia (SIVAM)," 7; and MOD, "Acórdão 194/2003—Plenário, Doc

AC-0194-07/03-P." During the mentioned period, no amendments or additional terms to Contract No. 001/95-CCSIVAM/Raytheon/Atech/Embraer were signed.

26. Federal Audit Court, *Audit Report: Implementation of the SIVAM Project*.
27. Pereira, *O CÉU É NOSSO!*, 9–12.
28. Pereira, 12.
29. This conception was presented by the Aerospace Defense Command (Comando de Defesa Aeroespacial Brasileiro, COMDABRA) in a presentation to students of the Specialized Tactics Instruction Group at the Natal Air Base on January 17, 2008. COMDABRA was replaced by the current Command of Aerospace Activities (COMAE) based in Brasilia.
30. Pereira, *O CÉU É NOSSO!*, 10–13.
31. Pereira, 12.
32. Civil Office, Decree 73,160; and Pereira, *O CÉU É NOSSO!*, 13–14.
33. Pereira, *O CÉU É NOSSO!*, 16.
34. Civil Office, Decree 6834.
35. MOD, "Ordem Do Dia Alusiva Ao 31 de Março de 1964."
36. MOD, "Política Nacional de Defesa e Estratégia Nacional de Defesa," 11.
37. For more on dissuasion as a concept, see Richard L. Kugler, "Dissuasion as a Strategic Concept," *Strategic Forum* no. 196 (December 2002): 1–8, https://www.files.ethz.ch/; Andrew F. Krepinevich Jr., "Dissuasion Strategy," report prepared for OSD/Net Assessment, December 2006, https://www.esd.whs.mil/; and MOD, "Política Nacional de Defesa e Estratégia Nacional de Defesa," 33.
38. MOD, "Política Nacional de Defesa e Estratégia Nacional de Defesa," 11.
39. Visacro, *A guerra na era da informação*.
40. MOD, "Política Nacional de Defesa e Estratégia Nacional de Defesa," 14.
41. Alvim, "How France Preserves and Explores Its Piece of the Amazon in French Guiana."
42. Adam et al., *RAPPORT FAIT AU NOM DE LA COMMISSION D'ENQUÊTE*.
43. Jayme, "FAB intercepta aeronave carregada com 165 kg de drogas."
44. Basseto, "Com radar, satélite e ataque a pistas, veja como a FAB defende e controla os céus da Amazônia."
45. Korybko, *The Law Of Hybrid War;* and Hoffman, *Conflict in the 21st Century*.
46. Santos, *A Evolução Do Poder Aéreo*.
47. Lonsdale, "Information Power." According to the author, the infosphere can be understood as the fifth strategic dimension, alongside the maritime, land, air, and space media. According to Lonsdale, this dimension of ethereal to polymorphic nature can be defined as the entity where information exists and flows.
48. MOD, "Política Nacional de Defesa e Estratégia Nacional de Defesa," 17.

Chapter 4

Brazilian Air Traffic Flow Management
An Integration Link in South America

Eduardo Sol Oliveira da Silva
Joaquim Lobo Júnior

Introduction

The air transport sector plays an important role in Brazil's economic activity and remains one of the fastest developing sectors in the world economy. In all regions of the globe, states depend on the aviation industry to stimulate economic growth and provide support for essential services to communities, especially the ones in inhospitable and remote regions—as with the indigenous tribes of the Amazon. In this context, civil aviation, by itself, can be seen as a significant contributing factor to the economic vitality of nations as well as to global economic vitality.[1]

One of the keys to protecting this economic vitality is the maintenance of a safe, efficient, and environmentally sustainable air navigation system available worldwide, nationally, and regionally. That requires implementing an air traffic management system that allows the maximum use of enhanced technical features. Thus, it is necessary that countries are developed in terms of aerospace structure, so that the well-being of their national bases, which involves issues ranging from the economy to air defense structure, is ensured.

In many places, demand often exceeds the system's available capacity to accommodate air traffic, resulting in negative, impacting consequences, not only for the aviation industry, but also for global economic health, hence the need to establish a balance between the demands and the capacity of airspace in these states. That is where air traffic flow management (ATFM) comes in.

Air traffic management (ATM) has become crucial for the various regions of the International Civil Aviation Organization (ICAO), including the South American Region. However, the economic and political characteristics of each nation and the possibility of investments in ATM are of paramount importance. So is interstate cooperation. How has Brazil helped, supported, or fostered the development and integration of air navigation service providers (ANSP) in South America through the ATFM system?

Background

In September 2003, at the ICAO's eleventh air navigation conference,[2] member states supported and approved the new *Caribbean/South American Air Traffic Flow Management Concept of Operation*. This document encouraged the implementation of a service management system that would allow the development of a continuous regional airspace through the application of a series of ATM functions.[3]

In accordance with ICAO's guiding principles regarding the facilitation of interregional harmonization, regional plans for the implementation of communication, navigation, surveillance (CNS) and ATM systems in the regions must be prepared according to the general profiles defined in the *2016–2030 Global Air Navigation Plan*.[4] After careful analysis, the guiding principles of this global plan were adopted by the CAR/SAM Regional Planning and Implementation Group (Grupo Regional de Planificación e Implementación del Caribe y América del Sur, GREPECAS), which incorporated the characteristics inherent to the Caribbean and South American regions (hereafter CAR/SAM), using as a foundation the definitions of homogeneous areas and main traffic flows. Homogeneous areas are air spaces with ATM requirements and similar degrees of complexity, while the main air traffic flows are air spaces in which there is a significant amount of air traffic demand.[5]

Considering the flow characteristic of the region in question, GREPECAS considered that the early implementation of ATFM would guarantee an optimum flow of air traffic in relation to or through some areas during periods when demand exceeds or is about to exceed available capacity of the air traffic control (ATC) system. Therefore, an ATFM system should reduce aircraft delays, both in flight and on the ground, and avoid overloading the system. The ATFM system will assist ATC in meeting its objectives and achieving more effective use of airspace and airport capacity. ATFM must also ensure that the safety of air operations is not compromised if critical levels of air traffic congestion occur and, at the same time, ensure that air traffic is effectively managed without applying unnecessary flow restrictions.[6]

The documents dealing with the concept of CAR/SAM air traffic flow management operations aim to describe a high level of service to be provided in these regions within a specific time horizon. They describe the status of the flow and what will be the future situation progressively achieved through a series of specific changes.

The operational concept described here reflects the expected order of events that may occur and should help and guide planners in the gradual design and development of the ATFM system, to provide safety and efficiency, in addition to ensuring an ideal air traffic flow to certain areas during periods when demand exceeds or is expected to exceed the available capacity of the ATC system.[7]

Brazilian Airspace and its Importance in South America

Brazil has continental scope. When it comes to air space jurisdiction, its dimension becomes even greater: there are 22 million square kilometers (8.5 million square miles) of continental and oceanic area. The extent of Brazilian airspace stands out compared to the areas of other South American countries. Its area, both territorial and spatial, is significant. Brazil also has the largest airport in South America, which is in the largest metropolis of the region—São Paulo.[8] This leads to the conclusion that Brazilian aviation affects the air movements of the entire region and, therefore, that Brazil plays an important role when it comes to aviation in the region. It is possible to infer that the development of national aviation benefits Brazil and its neighboring countries.

Because of this dimension and the growing demand, the Air Navigation Management Center (Centro de Gerenciamento da Navegação Aérea, CGNA), which was created in August 2005 by ordinance 1003/GC3, was activated in 2007 to manage national air flow demands. As a military organization subordinated to the Brazilian Airspace Control Department (Departamento de Controle do Espaço Aéreo, DECEA), CGNA aims to "allow, based on flight intentions, the harmonization of air traffic flow management, air space, and other activities related to air navigation, providing the operational management of actions processes of air traffic management (ATM) and related infrastructure processes, aiming at the sufficiency and quality of services provided under the Brazilian Airspace Control System."[9]

Thus, the development of Brazilian ATFM structures and aeronautical infrastructure represents a contribution to the development of South American aviation itself. Academic and aeronautical communities recognize this dimension of aviation development, as the excerpt below expresses (specifically referring to radar coverage over the entire Brazilian territory):

> In South America, Brazil . . . has this network of complete coverage of radar signals to support air navigation. Such technological advancement provides the Brazilian economy with an agile and safe means of transport. However, to be this way, it needs a considerable range of resources, both in its implementation and maintenance, as well as in its perennial need for updating. And if the number of commercial flights is increasing at high rates, as mentioned before, this entire infrastructure must keep pace with this growth, in order to continue providing the security and economy inherent in fast and reliable transport.[10]

According to the ATFM operational concept for the CAR/SAM regions,[11] Brazil accounted for 45 percent of the total movements at airports in the entire SAM region between 2002 and 2005. Projections made in 2013 anticipated that, after implementing ATFM from 2019 to 2021, Brazil would emerge as the largest industrial and economic power within the SAM region, with a probable increase in air traffic by 2030. Despite modest forecasts in relation to the increase of air traffic for the region, Brazil and Mexico still represent the most important domestic markets. Increased personal incomes and low-cost carriers (LCC) will drive future increases in traffic. The forecasted annual traffic growth is 5.9 percent by 2030 for the SAM region. For the entire CAR/SAM region, the forecasted growth is 74 percent in passenger transport.[12] These expectations confirm the propositions in previous chapters regarding Brazil's prominence in the South American continent's air traffic flow.

In summary, the increase in air traffic in Brazil impacts the flow in the region, and the increase of flow in South America impacts the air traffic flow in Brazil—especially when considering most flights destined for neighboring countries (originating mainly from Europe and Central Africa) pass through Brazilian airspace. The integration of

ANS providers of each country in South America is understandably mandatory for safe, orderly, and fast flow management.

The Caribbean and South American Monitoring Agency (CARSAMMA) is part of ICAO and monitors aircraft flight level maintenance performance, as part of a program for implementing the reduced vertical separation minimum.[13] To derive quantitative data on the impact of Brazilian airspace on the SAM region, CARSAMMA carried out a survey of overflight time data for the entire region. First, it analyzed the recorded flyover time from two consecutive years, 2016 and 2017, in each country. Next, it compared the overflight time of aircraft across the entire regional airspace, including Brazil. The analysis showed the average time of flight in Brazilian airspace and in the airspace of South America as a whole.[14]

Of the total flight times in South America, most took place within Brazilian territory. The total flight time in hours across the SAM region as a whole was 9 million hours. In South America, except Brazil, the total was 4.6 million hours, and in Brazil itself, the total was 4.5 million hours. Therefore, the proportion of total Brazilian flights in relation to all the flights in South America is 49 percent. In other words, half of all time flown in the SAM region is within Brazilian territory only. This study reveals the importance of Brazil in the SAM region and its significant impact on regional aviation.[15]

This data survey carried out by CARSAMMA concluded that flight time in Brazil is equivalent to 82 percent of the time when compared with flight time in other countries added together.

Brazilian Support for South American ATFM

It is not by chance that Brazil, through the Air Force Command, began a national development program for air traffic management technology. CGNA's creation in 2005 was one consequence of this evolution. Such evolution spurred development of the national aerospace equipment industry, with companies such as ATECH of Embraer group, which helped develop the Brazilian traffic management and control system, and Saipher, a Brazilian company specializing in the development and commercialization of software and solutions for ATC and ATM.[16]

Given these geopolitical assumptions of integration and in view of the various meetings with neighboring countries at the South American Implementation Group (SAMIG) as well as the demands arising from the ICAO guidelines for the SAM region—and with the prerogative of establishing an integration link and mutual aid for the development of ATFM in South America—the Air Force Command initiated the Tyr project through DECEA. The objective is to establish support to South American states to develop ATM activities. Tyr's main scope is to help promote ATC technical training in the region and cooperate with activities related to the security and efficiency of airspace control, such as restructuring, ATFM, and inspection in flight, among others. This project became an enterprise within the SIRIUS program, a program that aims to develop the national ATM system.[17]

The Tyr project has brought benefits to air circulation in the region, in addition to sharing with partner countries the systems that assist Brazil in the full implementation of ATC, ATFM, air space management, and more. As part of the training support, since 2015 CGNA has offered ATFM-related courses for countries in the SAM region, covering ATC capabilities, track capacity, flow management measures, and other ATFM concept resources.[18]

The purpose of these training courses is to provide students with technical knowledge of concepts related to the ATFM service, which is essential for the provision of the service with a focus on collaborative decision making (CDM). This process aims to improve the performance of the ATM system by harmonizing objectives and needs. As mentioned, Brazil trains ANS providers in the SAM region who are interested in implementing or improving the ATFM service in their respective states and seeks procedure and process harmonization with other countries in the region.

Between 2015 and 2018, CGNA had the opportunity to exchange knowledge about ATFM in the SAM region, offering training courses for operators to some South American countries, such as Argentina and Paraguay. In 2021, Brazil, along with other countries, proposed a joint effort to promote ATFM by activating or improving their respective ATFM cells. This would make it possible to increase the number of air traffic flow management positions (FMP) in the region and thus ensure greater integration between countries and the exchange of information between them, contributing to the growth of ATFM in South America as a whole.

In addition to the concepts to be taught at these meetings, Brazil offers a module about its flow management tool, the Integrated Air Movement and Management System (Sistema Integrado de Gerenciamento de Movimento Aéreo, SIGMA), which currently has a course with a web-based platform. In this system module, FMPs gain access to the movements included in the database. Users can see flight data originating in Brazilian airspace to the different airspaces and in the opposite direction. In addition, traffic originating outside Brazil but that does fly over some sectors of Brazilian airspace is also displayed and is thus available for air demand graphs and promoting safe, fast flow management, orderly and regionally harmonized, and interoperable, as recommended by ICAO.

With greater participation and integration of all FMPs in the region, air movements can be complemented by the respective demands presented by adjacent FMP cells. It will be possible to measure the expected demand of the airspaces of interest, evaluate the possible measures that will be applied, and coordinate their application time in a collaborative decision process that tends to grow and improve naturally.

It is important to clarify the need to analyze the capacities of each aerodrome of interest and airspace. Such research is crucial for the concept. Brazil has offered to train operators in the interested states and monitor their airspace capacities at the appropriate time.

Brazil's Support to South American Countries

On April 18, 2017, the technical cooperation agreement was signed between DECEA, the Argentine National Air Navigation Company, the National Directorate of Civil Aeronautics of Paraguay, and the National Directorate of Aviation Civil and Aeronautics Infrastructure of Uruguay.[19] This agreement develops collaboration and integration between the countries of the southern cone of the SAM region through the exchange of information; promotion of ATM technologies; consultancy; and specialized advice, teaching, and research. The agreement was one of the first actions taken by Brazil, via the Aeronautics Command through DECEA, to provide support to countries in the SAM region. In this way, Brazil reaches consonance with ICAO recommendations regarding cooperation efforts between countries for the development of air navigation.

Integration Agreement with Paraguay

On May 21, 2002, Brazil's president enacted Decree 4240, which deals with the "Mutual Cooperation Agreement between the Government of the Federative Republic of Brazil and the Government of the Republic of Paraguay to Combat the Traffic of Aircraft Involved in Illegal Transnational Activities, signed in Brasília, on February 10, 2000."[20]

The document details the mutual commitment between these countries that celebrates the use of the two characteristics present in the Air Force Command, namely the air force and aeronautics, to curb illicit traffic between states, especially drug trafficking. The agreement expands the margins of cooperation between the countries and their effectiveness and established the following activities by both signatory governments: (1) exchange of information to achieve the objectives of this agreement, (2) specialized technical or operational training, (3) provision of equipment or human resources to be employed in specific programs in the mentioned area, and (4) mutual technical assistance. In the agreement, the Brazilian government designated the Brazilian air force's general staff as coordinator for executing the agreement. The Brazilian and Paraguayan air forces oversaw establishing the work program.[21]

In August 2018, DECEA held its first meeting on the agreement and presented plans to develop the ATM system of Paraguay through Brazilian support.[22] The plans included the following:

a. the advisory for the implementation of an airfields and ground aids (AGA, aeródromos e auxílios terrestres) portal

b. assistance for the implementation of the drone/RPA portal

c. exchange of air traffic surveillance data between Brazil and Paraguay

d. basic courses on air traffic services (ATS) surveillance and techniques of ATS surveillance services in route and terminal areas

e. Team Resource Management course

f. Operational Safety Risk Management course

g. Operational Safety Management System (Sistema de Gerenciamento de Segurança Operacional, SGSO) course and organizations and entities providing the ANS

h. Track System Capacity course

i. ATC Sector Capacity course (air traffic control)

j. theoretical procedures for air navigation/operations courses

Brazil also will assist Paraguay in implementing the Guarani International Airport approach control (and possibly help implement SIGMA). All the equipment would be developed by Brazilian aerospace industries.[23]

In October 2018, DECEA sent a team of air traffic controllers to teach a practical instruction training course and instructor preparation internship for 22 students from the Asunción Control Tower, Approach; the Asunción Area Control Center (integrated); and Guarani Approach Control in addition to members of the administrative and electrical engineering sector of Paraguay. At the same time, DECEA started implementing the X-4000 platform, a console and radar synthesis system for the improvement of air traffic control in the country and, consequently, to strengthen the ATM system. This also helped reinforce cooperation and integration between countries in matters of defense based on common interests among participating states and noting the benefits and advantages arising from such partnerships between the forces.[24]

As part of this understanding, DECEA supported the implementation of the Guarani Approach Control (APP), which began, in April 2019, to operate the radar control of its terminal. That will contribute to the surveillance and security of operations in the Foz do Iguaçu region. The words of the then–general director of DECEA, Lieutenant Brigadier Jeferson Domingues de Freitas, express the main objective of these actions:

> Congratulations to all who promoted this work, as well as to our Paraguayan companions, who will receive instruction and training, which will allow the implementation of the APP Guarani and integrate, in the very near future, the CGNA, with an air traffic flow management cell in Asunción. This week we took a very important step to ensure safer, smoother, and more efficient flights for passengers from Brazil and Paraguay. DECEA—through CISCEA, CINDACTA [Centro Integrado de Defesa Aérea e Controle de tráfego Aéreo (Integrated center of Air Defense and Air Traffic Control)] II and DTCEA-FI [Destacamento de Controle do Espaço Aéreo (Unit of Air Traffic Control)], together with Paraguayan companions—took a very important step toward a more integrated future between the two countries.[25]

Lieutenant Brigadier Domingues points to regional integration that establishes an ATFM system that covers Paraguay, among other countries, bringing benefits to the entire SAM region.

In December of the same year, Brazil supported Paraguay through theoretical and practical instruction to civil and military air traffic controllers. Representatives from the control towers, approach control, and operational human resources areas received technical training. Such actions encourage and enable the development of a consistent, safe, and orderly air traffic flow management between Brazil and Paraguay.[26]

Integration Agreement with Argentina

The Performance Based Navigation Argentina project was created in 2016 on the same assumption of integration and cooperation with friendly nations, envisaging the engagement of Brazilian experts from the Aeronautical Cartography Institute to support the restructuring of airspace procedures that encompass Ministro Pistarini International Airport (Ezeiza), Aeroparque Jorge Newbery, and El Palomar airport, all serving the Buenos Aires province. This is one of the regions with the highest air traffic flow in South America. Brazilian specialists would use the computational tools acquired by the Argentine Air Navigation Company (Empresa Argentina de Navegación Aérea, EANA), such as Flight Procedure Design and Airspace Management (FPDAM) and Global Mapper, to carry out the task of preparing instrument flight procedure (IFP) charts for various airports in Argentine territory, in addition to training technicians from that country.[27]

In January 2017, DECEA brought in CGNA to support EANA in the development of an ATFM cell in the Ezeiza Area Control Center. Considering the fundamental principle of ATFM and the balance between available capacity and air traffic demand, CGNA established the following objectives: to create the operational concept of the Argentine ATFM; to teach the capacity calculation course for airspace sectors; and to analyze and calculate the capacity of the ATC sector at ACC Ezeiza, North Sector, Control Area (CTA) North, South Sector, CTA South, and Terminal Control Area Baires to carry out the initial analysis of capacity and demand in Argentine airspace. All of this was endorsed in a report from the Center for Studies on Air Navigation (Comissão de Estudos Relativos à Navegação Aérea Internacional, CERNAI) in July 2018.[28] Brazilian representatives taught ATFM technical courses using the methodology adopted in Brazil, which was focused on the following specific objectives: identify the normative basis of the ATC sector capacity study; understand the ATC sector capacity study; demonstrate the factors that contribute to imbalance in the ATC sectors; apply statistical concepts in analyzing ATC sector capacity; and justify the relationship between the ATC sector capacity measurement and operational safety for the ATC body. It was divided into practical and evaluation phases.[29]

To begin analyzing Argentine airspace capacity, information (details such as average time of controller-pilot communication, time spent within a specific portion of Argentine airspace, average time of coordination, and so on) was collected *in loco*, following the methodology adopted in Brazil, in the airspace of ACC Ezeiza and Ezeiza and Aeroparque airports. At the end, the Brazilian and Argentine teams met to present the results of the capacity measurement, in a CDM process.[30]

To formulate the Argentine ATFM operational concept, Brazilian representatives, with the support of CGNA specialists, worked with representatives of the Argentine ATM system to carry out the initial survey, analyze the specifics of the airspace, and begin planning ATFM implementation in the country. They drew from ICAO principles and recommendations (Global Air Navigation Plan, Global ATM Operational Concept, Collaborative ATFM, ATFM Operational Concept for SAM Region, among others) as well as from Brazilian ATFM experience and the performance of DECEA and CGNA since 2007.[31]

This activity was divided into three phases; the first was the formulation of an ATFM concept of operations. In the second phase, Argentinian ATC representatives would take a tactical ATFM course in Brazil. The third phase was the actual implementation of the Argentine ATFM, which took place in May 2018 with the first activation of a Flow Management Unit (FMU) in Argentina under the coordination of Brazil.

Most recently, in December 2018, Brazil signed a search and rescue knowledge-transfer agreement with Argentina for the area.

Agreement with Bolivia

In December 2018, the Brazilian Aeronautical Technical Mission in Bolivia (Missão Técnica Aeronáutica Brasileira [MTAB]-Bolivia) was established to provide administrative, technical, and operational support for the integration of air navigation services to the countries of South America, to support Bolivia's Comando de Seguridad y Defensa del Espacio Aéreo (COSDEA) in matters related to air defense and air traffic control systems.[32]

Their objective was to share their lessons learned to help Bolivia's ATFM development. Many training courses in different areas have already occurred, as have the low-cost radar simulator and several operational training courses about air traffic control.

Brazil's participation in Bolivia's airspace management process confirms the importance of cooperation and integration among the countries of South America. Brazil is one of the main links within this context for the SAM region, and its contributions directly influence the development of aviation in the area, providing airspace in the southern cone and making it more homogeneous and interoperable.

Conclusion

This chapter sought to highlight the important role of Brazil within the context of South American ATFM integration to answer the following question: how has Brazil helped, supported, or fostered the development and integration of air traffic navigation service providers in South America? The airspace control system and the aerospace defense system perform both control and defense for SAM airspace. DECEA developed the Air Navigation Management Center, directly responsible for managing Brazilian air traffic flow and supporting the airflow management of some countries in South America. This was one of the most significant airspace infrastructure developments that allowed, consequently, the development of civil aviation.

The peaceful and proactive relationships with neighboring nations, guidance and statutes in Brazilian law from the federal constitution to the National Defense Plan, and DECEA support policy and continental ATM integration make Brazil an important player in the region's aeronautical community.

Notes

1. International Civil Aviation Organization (ICAO), "Global Air Traffic Management Operational Concept."
2. ICAO, "CAR/SAM Seminar in Preparation of Eleventh Air Navigation Conference (AN-Conf/11)."
3. ICAO, *Caribbean/South American Air Traffic Flow Management.*
4. ICAO, *2016–2030 Global Air Navigation Plan.*

5. ICAO, *Caribbean/South American Air Traffic Flow Management*.
6. ICAO.
7. ICAO.
8. Santos, Neiva, Pizolato, and Spigolon. "Conceito cidade aeroporto: Guarulhos como aerotropolis."
9. Ministry of Defense (MOD) and Air Force Command, Decree 1547/GC3.
10. Porto et al., "O Meio de Transporte Aéreo como Apoio Logístico."
11. ICAO, *Caribbean/South American Air Traffic Flow Management*.
12. ICAO, *Global Air Transport Outlook to 2030 and Trends to 2040*, 12.
13. Air Force Agency, "Portal Operacional ATFM-CGNA."
14. Lôbo Júnior, "O gerenciamento de fluxo de tráfego aéreo."
15. Lôbo Júnior.
16. Lôbo Júnior.
17. Lôbo Júnior.
18. Lôbo Júnior.
19. Meireles, "Foz Do Iguaçu Sedia a 1a Reunião."
20. Civil Office, Decree 4240.
21. Civil Office.
22. Meireles, "Brasil Coopera Com Paraguai."
23. Meireles.
24. Meireles; and Guimarães, "DECEA Inicia Implantação Da Plataforma X-4000 No Paraguai."
25. MOD and Air Force Command, Decree 2146/GC3, 15.
26. MOD and Air Force Command, Decree 126/SDAD.
27. Meireles, "Instituto de Cartografia Aeronáutica Colabora."
28. ICAO, *Proyecto Regional RLA/06/901*.
29. Lôbo Júnior, "O Gerenciamento de Fluxo de Tráfego Aéreo."
30. Lôbo Júnior.
31. Lôbo Júnior.
32. MOD and Air Force Command, Decree 2146/GC3.

Part 2

Aerospace Power and Contemporary Issues

Introduction to Part 2

Part 2 of this book covers contemporary issues related to Brazil's aerospace power arm as it intersects with national policies and military doctrine, international humanitarian law and operations (and the attendant interoperability concerns), and geopolitics writ large.

Brazil has been a partner nation in missions worldwide, for example with the United Nations Stabilization Mission in Haiti (Mission des Nations Unies pour la stabilisation en Haïti, MINUSTAH), requiring consideration of its laws and procedures as Brazilian armed forces operate on an international scale. We examine this from a point of view not exhaustively explored in other studies: the cultural intelligence approach. Geopolitics is also observed from the aerospace point of view, reinforcing the idea of integration between the air and space domains, with a new national approach of an aerospace geopolitics.

These issues reveal the phenomenon of the modernization of warfare in the Brazilian context. This implied the establishment of strategic sectors of national defense, focused on the expansion of nuclear, cybernetic, and space capabilities. The use of these new mechanisms in Brazil must comply with international humanitarian law (IHL). In this context, IHL application in issues considered strategic for Brazilian national defense is a central current issue that should be discussed given Brazil's participation in the UN Security Council and in several peace operations. Even internally, national defense documents maintain the pillars of the Brazilian traditional vision, contrary to the war of conquest, which defends peace in the resolution of conflicts and international cooperation as an instrument to achieve these objectives, as well as respect for IHL in military operations.

In the case of peace operations, humanitarian support and security activities seem to be related to global stability, and aerospace power can be considered an essential element in the development of UN peace and security activities. Brazil's aerospace power adopts a professional and rigid attitude to make humanitarian aviation a reality, as current operating conditions are close to the ones observed in peace operations. Doctrinal precepts and legal documents reinforce the demand for the capacity to act in the international arena, with an objective to contribute to world peace and to the fulfillment of international commitments. There is even a joint center that carries out doctrinal training to prepare military personnel to act in the peace operations environment. Consequently, the condition of operational readiness is established, which mutually favors the cooperative character of the country's international relations and the exploitation of the dissuasion principle through the projection capacity of aerospace power.

There must be a consensus among the armed forces on the question of interoperability for capabilities like these to be effective. This is a current issue, especially when discussing the reduction of the military budget. But this is not the only possible approach to the jointness problem. One aspect that has been little studied, especially in Brazil, is the human factor in this equation. There are increasing discussions about leadership, communication, and interpersonal relationships, which aim to reach hearts and minds by building on the concept of interoperability. It is relevant to consider the importance of the cultural intelligence approach in the current context of stimulating joint operations in Brazil. Both organizational and doctrinal cultural peculiarities bring the possibility of improving interoperability. The current scenarios of joint employment demand flexibility and versatility, deepening the discussion of the importance of contemplating—besides the doctrinal component—the human and cultural components in joint operations. It is a matter of making the joint process more effective, operating in the field of human resources to achieve the maximum possible synergic effect.

Dealing with issues such as human resources is of fundamental importance to aerospace power. Although technology is a fundamental component in this area, it is effectively the people who drive systems and equipment, and there are several military and political figures who reinforce the priority given to the human factor in conflicts. At a global level, the SARS-COv-2 pandemic provoked a reorganization of international geopolitics to the extent that its impact on the human factor was resounding. It affected economic, cultural, political, social, and labor relations (employer and worker) as well as the ways of performing tasks, intensifying remote work. For this reason and in line with the demand for analysis under a cultural focus, a scientific dialogue between geopolitics, culture, and the impacts on aerospace power, especially on the issue of cyberspace, becomes necessary. In fact, all these areas of knowledge are included in three major epistemological flows, which will result in the reconfiguration of the postpandemic world, whose projection points to an intense dispute for technological innovation in cyberspace. Therefore, it is urgent to investigate the role of aerospace power considering current geopolitical tensions.

Aerospace geopolitics is related to both science and aerospace power. The scientific relation covers a wide spectrum of phenomena in economics (e.g., air market regulation or asteroid resources exploration), politics (e.g., weaponization of outer space), and culture (e.g., ideological discourse around space programs). In this new environment they have become decisive for mankind and for shaping power relations, defining territories, creating legal frameworks, making sovereignty claims, and settling disputes that go beyond the classical approach of geopolitics on the surface. This new geographical dimension to the study of geopolitics is about revealing an aerospace environment, defined as a geographic domain formed by the combination of airspace and outer space, in which geopolitical relations are established. Geopolitical analysis in this physical domain points to evidence in the fields of geographic epistemology, with political, economic, technological, and ideological variables. Aerospace power is reflected in these geopolitics, emphasizing its military nature, the possibility of exploiting natural resources, the growing importance of air transport, access to new communication and information technologies, economic development generated by the aerospace industry, the presence of ideologies, and even the possibility of man making better use of the atmosphere and exploring, or inhabiting, celestial bodies. Thus, the aerospace environment is a new domain for geopolitical science and a stage for national defense and development strategies, reinforcing the relevant role of aerospace power for society.

Again, it is reaffirmed that the concept of aerospace power in Brazil has a comprehensive and multidimensional view. The chapters in part 2 highlight current themes of this concept. First, the emphasis that Brazil—in particular the Brazilian air force—has placed on participating in United Nations humanitarian operations; to this end, adherence to the principles of IHL has been a basic condition for Brazilian action. Secondly, aerospace power is seen as a major component that combines forces of various natures and, therefore, demands coherent interpretations of the idea of jointness and interoperability, including the cultural aspect, which values the human factor. Finally, there is the interpretative bias of geopolitics as a science that is intrinsically related to aerospace power, in the form of either reinterpretation of the society of postpandemic international relations or constitution of an environment or geographic domain in which phenomena are established, such as, but not limited by, classical geopolitics, a geopolitics of aerospace power, or an aerospace geopolitics. In this sense, the peculiarity of Brazilian aerospace strategic thinking is again observed.

Chapter 5

Brazil and International Humanitarian Law
Application in Strategic Sectors for National Defense

Luciano Vaz Ferreira
Carlos Alberto Leite da Silva
Luís Eduardo Pombo Celles Cordeiro

Introduction

"All is fair in love and war." This is a widespread popular saying to justify situations in which there are simply no rules. However, any student of the art of war knows that at least half of this sentence incurs a crass error: in war, indeed, there are rules.

Starting from the idea that states will always be subject to the threat of conflicts, it was understood that allowing total freedom of action in war would be disadvantageous even for the winners, since the effects of a given event could overcome the gains obtained on battlefields. In this context, states decided to create codes of conduct to minimize the side effects of the struggle so that only the combatants were affected and only military objectives were attacked. Over time, such writings became codified and adopted as part of international humanitarian law (IHL).

Obviously, Brazil was not immune to such events: it was both influenced by and influenced these changes. Despite its scarce participation in wars, its role as a diplomatic agent in the international field, both in the creation of the United Nations (UN) and in supporting UN peacekeeping operations, spurred internal debate about IHL, thus influencing even the preparation and the employment of Brazilian armed forces.

The Brazilian government's point of view in relation to the modernization of war necessitates strategic sectors for national defense focused on the expansion of nuclear, cyber, and space capabilities. However, the use of these capabilities must take place in line with the international legal system, IHL in particular. This chapter discusses the application of IHL in issues considered strategic for Brazil's national defense. A literature review identifies the main points of tension and elements for debate. The chapter is divided into three sections: the first is an introduction to IHL; the second deals with the interrelationships between diplomacy, defense, and international law from a Brazilian perspective; and the third is about the application of IHL in national strategic sectors.

The Development of International Humanitarian Law

Since ancient times, relations between human social groups (from rudimentary tribes to nations) have been marked by violence. However, evidence shows that war, although brutal, always involved some kind of legal limitation for conduct considered beyond the pale among belligerents.[1] Nations began to craft laws over time, and those laws formed a collection now referred to as public international law, an important instrument to ensure the sovereignty of states and induce certain standards of conduct. Legal restrictions in conducting hostilities developed in response to the continuous strategic and technological innovations that elevated the destructive character of war, either by causing unnecessary suffering to combatants or by targeting defenseless people. These standards became increasingly recognized by nations, especially by signing international treaties (bilateral or multilateral), which have proliferated in the last two centuries.

Nineteenth-century conflicts yielded significant experiences that led to the application of international law. In 1859, during a business trip, Swiss businessman Henri Dunant witnessed the Battle of Solferino, involving French, Sardinian-Piedmontese, and Austrian troops. In a single day, about 40,000 soldiers were injured or died, mainly due to the lack of medical aid on the battlefield. Shocked, Dunant wrote a book reporting his testimony (*A Memory of Solferino*) and began to spread the need for regulation of war throughout Europe. In the same year, along with other notables from Swiss society, Dunant founded the "International Committee for Relief to the Wounded," a private entity created with the goal of disseminating Dunant's ideas. In a short time, he invited world leaders to discuss the possibility of drafting an international document to protect the wounded in battle. As a result, in 1864, after the First Geneva Conference, the "Geneva Convention for the Amelioration of the Condition of the Wounded in Armies in the Field" was created. It provided for the protection of medical personnel on the battlefield and assistance and repatriation of wounded enemy combatants. The committee Dunant created was renamed the "International Committee of the Red Cross" in 1876 and still plays an important role today.[2]

Across the Atlantic, in 1863, US President Abraham Lincoln enacted the Lieber Code. It was a set of rules of conduct for the armed forces of the United States and used during the Civil War. Among other things, it banned torture, the use of poisons, and slaughtering captured enemy forces.[3]

From this beginning, additional international treaties followed one after the other until the Second World War.[4] After World War II and a changed international political landscape, the humanitarian cause gained new momentum. The UN Charter and Universal Declaration of Human Rights represented attempts to establish an international social contract based on the primacy of peace and the promotion of human rights.[5] The trials of war criminals by the Nuremberg and Tokyo Courts in 1948 opened the possibility of individual punishment for those who violated the law of armed conflict.[6]

The Fourth Geneva Conference, held in 1949, represented a milestone for this new period. The attendees reviewed and updated previous initiatives and systematized the fourth convention's additions; together these became the backbone of the legal limits in the conduct of hostilities.[7] The conventions were complemented by three Additional Protocols (signed in 1977 and 2005),[8] in addition to several instruments signed later.[9] The combination of these treaties and related customary international norms form international humanitarian law.

IHL is concerned with three fundamental questions: (1) who and what may be the target of a military operation, (2) what methods and means can be used in these operations, and (3) what treatment should be given to those captured during the conflict. The texts of IHL conventions are more detailed than this current analysis can encompass, thus, we investigate the possibilities of applying IHL tenets that are now considered basic parts of international custom.

In general, IHL attempts to balance military needs and human dignity. To subdue an adversary, it may be militarily necessary to cause death, injury, or destruction, in addition to imposing more severe security measures than in peacetime. However, this does not mean that the conduct of hostilities may occur unrestrained and without imposing humanitarian constraints.[10]

This balance is guided by humanitarian principles such as the following:

1. distinction: it establishes that the legitimate targets in a conflict must be strictly combatants, protecting noncombatants (such as the civilian population) and civilian objects;
2. precaution: duty to avoid or minimize possible collateral damage to civilians;
3. proportionality: related to the need to measure the concrete objectives and the anticipated military advantage with the probable civilian harm, when it is not possible to avoid them;
4. unnecessary suffering: to prohibit means and methods of warfare that may cause unnecessary suffering and superfluous damage to combatants;
5. humane treatment: to maintain humane treatment that should be granted to all persons in the power of the enemy, regardless of their previous status or function.[11]

It is noteworthy that in the post–World War II world, the term "armed conflict" has taken precedence over "war." This is because of the 190 conflicts that have erupted around the world since 1945, the parties involved considered the event to be an act of belligerence or of war in only 19. This highlights the need for IHL adoption for conflicts below the level of formally declared war.[12]

Distinctions have been delineated between international armed conflict (IAC)—direct military confrontation between high contracting parties (states)—and noninternational armed conflict (NIAC), confrontation between the military forces of a state and one or more nongovernmental armed groups that dominate a part of the territory where they exercise effective control or between nongovernmental groups. Currently, these are the only legally recognized forms of armed conflict and, therefore, where the application of IHL is mandatory (IAC and NIAC are not considered internal conflicts, disturbances, or tensions, even if armed forces are used).[13]

Recent technological advances that include autonomous and remote-controlled weapon systems, cyberattacks, and military use of outer space seem to test IHL application limits.[14] The asynchrony between technological changes and the law (including at the international level) is typical, since law updates move more slowly than society does.[15]

In accordance with Article 36 of the First Additional Protocol to the Geneva Conventions, the state that develops, acquires, or adopts new weapons or methods of combat is obliged to verify that their employment is consistent with IHL. This obligation can be met through a commission or advisory body to the ministry of defense and helps establish a culture of prior verification of the legality of the means and methods of combat.[16]

Diplomacy, International Law, and the Brazilian Armed Forces

How does this set of international norms relate to Brazil's foreign policy, legal order, and armed forces? Traditionally, Brazilian foreign policy—which includes its defense policy—has focused on "peace, Law, moderation and transaction."[17] Such ideals have been shaped since the beginning of the formation of the Brazilian state and its independent diplomacy. The state council of the Brazilian Empire, one of the first consultancy bodies in national diplomacy, drafted its opinions advising that Brazilian foreign policy should be based on the defense of international law. The inspiration came mainly from the influence of the law schools of Olinda and São Paulo, training centers for intellectuals who made up the state and diplomatic staff of the Brazilian government.[18]

After the promulgation of the republic, Brazil maintained this approach. During the beginning of the republican period, Brazil participated in the Second Hague Convention of 1907, mainly defending the prohibition of war for debt collection between states and the improvement of international arbitration mechanisms. During the administration of José Paranhos, Baron of Rio Branco (1902–1912), the biggest name in national diplomacy, parties resolved disputes in Brazil which involved the definition of Brazilian borders through negotiations based on international law and not by force, despite being the country with the greatest military contingent in South America. Herewith, it was argued that the country that historically occupies a territory has the right to annex it (*uti possidetis*).[19] The country actively participated in peace conferences at the end of the two Great Wars and was a founding member of both the League of Nations and the United Nations.[20]

An important aspect of Brazil's performance on the international stage, one that demonstrates its defense of peace and respect for international institutions, is its participation in the UN Security Council and in UN Peacekeeping operations (PKO). Brazil has been elected to a nonpermanent seat on the Security Council 11 times, being one of the member states that has participated most in the institution, second only to Japan.[21] In 1956, Brazil sent military personnel for the first time to a PKO on the Suez Canal. The UN has conducted more than 70 peacekeeping missions, and Brazil has collaborated on more than half of them. Noteworthy among the missions are Brazil's participation in the missions in Haiti (MINUSTAH), Democratic Republic of Congo (Missão Nações Unidas de Estabilização no Congo, MONUSCO) and Lebanon (United Nations Interim Force in Lebanon, UNIFIL), crises in which it exercised military command.

Brazil has a long tradition of expressly guaranteeing in its fundamental rules the prohibition of the war of conquest, the defense of peace, and international cooperation, as can be seen in the texts in the Constitutions of 1891,[22] 1934,[23] 1946,[24] 1967,[25] and 1988. The current Constitution of the Federative Republic of Brazil (1988) includes as principles that govern the country performance at the international level (art. 4): national independence, prevalence of human rights, self-determination of the peoples, nonintervention, equality among the states, defense of peace, peaceful settlement of conflicts, repudiation of terrorism and racism, and cooperation among peoples for the progress of mankind. The 1988 federal constitution innovated by including the prevalence of human rights as a fundamental principle of the Brazilian nation.[26] The principles of international relations in its constitution certainly oblige the Brazilian state to respect international law and IHL in its foreign policy.

According to the constitution, the Brazilian armed forces (army, navy, and air force) are intended to defend the country and guarantee constitutional powers, law, and order (art. 142). It observes the defensive character of the mandate of the military institutions, which obviously are also subject to the principles of international relations and compliance with IHL.

Currently, the main documents on national defense in Brazil are the *National Defense Policy* (*Política Nacional de Defesa*, PND), *National Defense Strategy* (*Estratégia Nacional de Defesa*, END), and the *Defense White Paper*.[27] The greatest merit of these instruments is the expansion of participation in the preparation of national defense planning, a discussion that encompasses not only military institutions but also different sectors of society. The current documents are from 2012, but in 2020 a new update went to the National Congress for approval.

The *PND* 2012 and *END* 2012 defend a peaceful international order based on democracy; multilateralism; cooperation; proscription of chemical, biological, and nuclear weapons; and strengthening principles enshrined in international law. The defense of peace and international cooperation are elements of great importance. The proposed 2020 *PND* and *END* follow the same line in defense of the Brazilian constitution and international law.

In 2010, the Inter-American Court of Human Rights, linked to the Organization of American States, condemned the Brazilian government for human rights violations between 1972 and 1975, during the Araguaia Guerrilla War. Among other obligations, the international body demanded the establishment of a permanent and mandatory training program in the protection of human rights aimed at all levels of the armed forces. The training should deal with instructions on "serious violations of human rights and military criminal jurisdiction" and "international human rights obligations," which include the dissemination and study of IHL.[28] IHL is part of basic courses for soldiers and specialists; officer training; specialized training for intermediate, superior, and general officers; courses in general staff and senior studies; and courses for military personnel participating in UN Peacekeeping Operations.

In 2011, the Brazilian Ministry of Defense published a document about IHL, with the objective of consolidating a normative instrument for the armed forces for dissemination, study, and consultation on the subject—*Manual de Emprego Do Direito Internacional Dos Conflitos Armados (DICA) Nas Forças Armadas*, MD34-M-03 (International armed conflict law employment handbook in the armed forces). It must be used in planning and employment, whether singular or joint, of the operational commands activated in IAC and NIAC. According to the document, the "Brazilian State has a significant predisposition to abide by the rules of International Law." The manual reproduces the principles and norms of IHL, emphasizing the obligation of strict compliance by the Brazilian armed forces when involved in an armed conflict, regardless of the existence of declared war.[29]

Strategic Sectors and the Application of IHL in the Brazilian Context

According to the Geneva Conventions and the military necessity perspective, an attack must be strictly limited to military objectives whose destruction (total or partial), capture, or neutralization offers military advantage concerning the adversary. There is a duty to safeguard the lives of those who are not directly related to the conflict. Collateral damage must be directly linked to the conflict context. Consequently, the planning and conduct of aerospace operations must comply with the principles of IHL, avoiding airstrikes on hospitals, schools, refugee camps, and other civilian installations. The role of the legal advisor is critical in the process of target selection for the use of airpower. The aim is to advise the commanding officer on the best options to satisfy the military objective within IHL requirements.

Currently, modern doctrine defends a multidomain approach, with airpower integrated not only with land and sea domains but also with cyber and space. Using cyber and electronic warfare to undermine the enemy's capacities is a reality on the battlefield, allowing better results for an airstrike. The employment of remote and autonomous air systems, such as unmanned aerial vehicles, depends on military cyber capabilities. On the other hand, the lines between air and space are blurred: Target acquisition technology relies on satellites in space; intercontinental ballistic missiles travel through air and space; suborbital flight is a tangible possibility. Therefore, the planning and execution of air operations must observe all possible variables brought by the new domains and their effects on IHL.

In the past few decades, Brazilian national defense planning has emphasized new military technologies. The 2005 *PND* highlights, albeit briefly, the need to reduce the vulnerability to cyberattacks. The 2007 *END* introduces the importance of developing national military technology in sectors considered to be "strategic": nuclear, space, and cyber. The 2012 *END* details the three sectors in a separate chapter.

In the 2012 *END*, the nuclear sector is directed toward the use of nuclear energy and the consolidation of a nuclear-powered submarine program. Currently, the Brazilian navy and the government of France are conducting the Submarine Development Program (Programa de Desenvolvimento de Submarinos, PROSUB), a partnership to build the first Brazilian nuclear submarine.

The strategy for the space sector focuses on the design and manufacture of satellite launch vehicles, geostationary satellites for telecommunications and remote sensing, communication technologies, command and control from satellites, and geolocation technology. The Brazilian air force is coordinating this strategy. In 2017, Brazil launched the Geostationary Satellite for Defense and Strategic Communications (Satélite Geoestacionário de Defesa e Comunicações, SGDC-1) in collaboration with France. On that occasion, it also created the Space Operations Center (Centro de Operações Espaciais, COPE). The goal is to provide security in strategic communications from government and military organizations through the exclusive operation of a geostationary satellite.

The cyber sector is also considered strategic for national defense. In 2010, Brazil created the Cyber Defense Center, associated with the Brazilian army, with the objective of expanding capabilities in the area. In 2014, the implementation of the Joint Cyber Defense Command began. In addition to developing the country's general capabilities (including the civilian sector), the objective is to contribute to national cybersecurity, in order to develop the capacity, preparation, and employment of operational and strategic cyber powers, in favor of joint operations and protection of strategic infrastructures.[30]

In this scenario, it is imperative to question what Brazil's possible limits are in the use of new technologies (nuclear, space, and cyber), observing its diplomatic-military tradition. Brazil endorses the defense of peace and respect for international law as important values for the nation, evident in the practice of its foreign policy, in its constitution, in the formation of its armed forces, and in the planning of its national defense. Thus, there is no doubt that any military operation of the Brazilian armed forces must strictly follow the provisions of IHL norms. The challenge is to try to interpret and overcome the possible existing gaps involving the newer technologies, since IHL predates them.

Regarding nuclear technology, there are no problems. Despite the existence of discussions about the possibility of development in the past, Brazil does not have nuclear weapons and does not intend to develop them. From a legal perspective, the country is a signatory to the 1967 Tlatelolco Regional Treaty, the 1968 Nuclear Weapons Non-Proliferation Treaty, and, most recently, the 2021 Nuclear Weapons Ban Treaty. The development of the Brazilian nuclear submarine does not seem to have any impact on the interpretation and application of IHL in the Brazilian context.

Regarding space, the main current international discussion concerns its militarization and weaponization. Space operations have their own self-contained regime, independent of IHL, consisting mainly of the 1967 Outer Space Treaty. There are three interpretations for article 4 of the treaty: (1) any military activity in space would be prohibited;[31] (2) military activities would be allowed, as long as they do not imply the use of weapons; and (3) military activities would be permitted, including the use of conventional weapons in space, while nuclear and mass destruction weapons would be prohibited.[32] The defenders of this last thesis understand that the "peaceful purpose" of the use of space must be understood as a "non-aggressive purpose,"[33] thus allowing defensive operations. In this case, it would be possible to use antisatellite weapons, technology currently used by the US, China, Russia, and India.

Discussions on the establishment of a new international treaty banning any space weapons have been around since the 1980s, at the Conference on Disarmament (Proposed Prevention of an Arms Race in Space, PAROS). In 2008, China and Russia jointly presented a new proposal (Prevention of the Placement of Weapons in the Outer Space Treaty, PPWT);[34] however, negotiations have not progressed so far.[35]

The application of IHL in conflicts over space depends on the means and methods of attack. In an attack on Earth from space (Space–Earth) the concern is to affect the civilian population, a similar situation to when airpower is used, with the use of missiles and aircraft. The possibility of using tungsten rods on satellites to be launched toward Earth and cause kinetic impact has already been studied.[36] The problem would be the lack of precision in this type of attack, something that would inevitably violate IHL.

In attacks departing from Earth toward space (Earth–Space) or departing from space to reach space artifacts (Space–Space), the current concern would not lie in the protection of people themselves, due to the rare human presence in space (except for the continued presence on the International Space Station and future manned missions). The disquiet concerns the attack on civilian objects in orbit, which may represent critical elements of infrastructure, such as telecommunication satellites and geolocation.

In principle, these devices cannot be attacked, either from Earth or from a co-orbital enemy satellite. The problem is that most satellites are dual-use—that is, they have both military and civilian purposes, such as GPS. In this situation, IHL would only allow a military operation based on the presence of two cumulative criteria: first, when the item effectively contributes to enemy military action, due to its nature, location, purpose, and current use; and second, if the attack on the device represents a definite military advantage, that is, manifests itself in a concrete and perceived way.[37]

Although space is part of the *END*, Brazil has not demonstrated interest in developing weapons in this area, even in a defensive way. In 2012, Brazil publicly defended the PPWT proposal in international forums and expressly manifested its opposition to the use of space weapons, including for self-defense. In 2015, the country supported the UN General Assembly Resolution on the resumption of PPWT negotiations and signed the political commitment that "it will not be the first to place weapons in space," together with Argentina, Armenia, Belarus, Cuba, Indonesia, Kazakhstan, Kyrgyzstan, Russia, Sri Lanka, Tajikistan, and Indonesia. The recent launch of the dual geostationary satellite seems to demonstrate that Brazil accepts the military use of space if it is limited to communication activities and does not involve armaments.

Regarding the cyber sector, in 2014, the Military Cyber Defense Doctrine (MD31-M-07) of the Ministry of Defense was approved. In the document, cyber warfare is defined as "offensive and defensive use of information and information systems to deny, exploit, corrupt, degrade or destroy the adversary's Command and Control capabilities, in the context of military planning at the operational level or tactical or military operation." Its use in the Brazilian case must be "in an environment of crisis and conflict, in support of the military operation."[38]

Three possibilities for using cyber actions are mentioned: (1) cyberattack: actions to interrupt, deny, degrade, corrupt, or destroy information or computer systems stored in the opponent's computer and communications devices and networks; (2) cyber protection: actions to neutralize attacks and cyber exploitation against computer devices and national networks, being a permanent activity; and (3) cyber exploration: search or collection actions, to obtain situational awareness in the cyber environment, aiming at producing knowledge or identifying cyber vulnerabilities.

Operations are based on four principles: (1) principle of effect: actions in cyberspace must produce effects that translate into strategic, operational, or tactical advantage that affect the real world, even if these effects are not kinetic; (2) principle of concealment: active measures must be adopted to conceal itself in cyberspace, making it difficult to trace offensive and exploratory cyber actions, to mask the authorship and the point of origin of these actions; (3) principle of traceability: effective measures must be adopted to detect offensive and exploratory cyber actions against friendly information and communication technology systems; and (4) principle of adaptability, the ability of cyber defense to adapt to the changing characteristics of cyberspace.

Compared with the nuclear and space sectors, the cyber sector seems to be the one in which Brazil has the greatest propensity to use military force. It is also precisely in this area where there is the greatest risk of producing possible IHL violations. Therefore, Brazilian doctrine in this area must always be adapted to such international standards.

In military cyber defense doctrine, the principle of effect must consider not only the desired advantage but also the legitimacy of the target based on IHL, whether the effects are kinetic or not. In exercising the principle of concealment, it is mandatory to maintain a control and record system for operations so that, if necessary, accountability for actions that infringe on IHL is carried out. The principle of traceability, which aims to monitor cyberspace, must be exercised in accordance with the principles of distinction, precaution, and proportionality to avoid civil damage.

Finally, the need for Brazil to ensure compliance with article 36 of the First Additional Protocol to the Geneva Conventions is reinforced. In this context, the country must periodically assess how new means and methods of combat linked to new technologies—which currently include the cyber and space sectors—align with IHL.

Conclusion

IHL was created to mitigate the effects of war. From this idea emerged legal instruments that were accepted in the international community since all the actors involved would be favored. The means and methods of combat were limited to avoid unnecessary suffering and superfluous damage, distinctions made between people and places that could be the target of military attacks and those who should be protected, and guidelines established for how prisoners of war should be treated.

In this continuous process of developing the debate, the contribution of Brazil, a country which historically defends IHL and other international standards, can be highlighted, like its participation in the UN Security Council and in peacekeeping operations. Also noteworthy is the insertion of IHL as a mandatory subject in Brazilian professional military education, a project of great importance.

At the political level, current national defense documents maintain the pillars of the Brazilian traditional view, which opposes wars of conquest, favoring instead the defense of peace in conflict resolution and of international cooperation as a tool to achieve these objectives, as well as respect for IHL in military operations. The main challenge to the strategic sectors foreseen in *PND* face is interpreting century-old concepts for application in contemporary military actions in fields as complex and dynamic as the nuclear, space, and cyber sectors.

Notes

1. There are records of limitations on the conduct of the belligerents in religious and legal instruments in Mesopotamia, Ancient Egypt, Palestine, Ancient Greece, and the Roman Empire. In the Middle Ages, the idea that war should have a limitation marked the thinking of Christian philosophers, such as Saint Augustine and Saint Thomas Aquinas. During the Renaissance, the doctrine of just war continued with the writings of Hugo Grotius, Alberico Gentili, and Emer de Vatel. The formation of a "perpetual peace confederation," which implied the limitation and prohibition of war, was suggested by the Abbot de Saint-Pierre and Immanuel Kant. Ferreira, *Direito Internacional da Guerra*.

2. Melzer, *International Humanitarian Law*, 35.

3. Melzer, 35.

4. St. Petersburg Declaration of 1868, which prohibits the use of explosive and flammable projectiles in wartime; Washington Treaty of 1871, requiring obligations of neutral states and times of war; The Hague Convention of 1899, concerning the laws and uses of land warfare; The Hague Convention of 1899, for the adaptation to the maritime war of the Geneva principles of 1864; 1906 Geneva Convention, to improve the fortunes of the wounded or sick in campaign exercises; The Hague Conventions of 1907, referring to land and sea warfare, the right and duties of neutrality, and the prohibition of launching projectiles from balloons. In the interwar period, the 1929 Convention on the Treatment of Prisoners of War stands out.

5. Ferrajoli, *A Soberania No Mundo Moderno*.

6. Gonçalves, *TRIBUNAL DE NUREMBERG 1945–1946*.

7. The First Geneva Convention of 1949 protects wounded and sick soldiers in land war; the Second Convention protects wounded, sick, and shipwrecked soldiers and sailors during maritime war; the Third Convention is aimed at prisoners of war; and the Fourth Convention refers to the protection of civilians. International Committee of the Red Cross (ICRC), "Geneva Conventions of 1949, Additional Protocols and Their Commentaries."

8. ICRC, "Geneva Conventions of 1949, Additional Protocols and Their Commentaries." The First Additional Protocol of 1977 deals with the protection of victims

of international conflicts; Additional Protocol II of 1977 refers to the protection of victims in non-international conflicts; Additional Protocol III of 2005 is about the adoption of an additional distinctive emblem.

9. The Hague Convention on the Protection of Cultural Property in the Event of 1954 Armed Conflict; Convention on the prohibition of the development, production and storage of bacteriological and toxic weapons, in 1972; 1980 Conventional Weapons Convention and its protocols prohibiting the use of nonlocatable fragments, mines, tramp weapons and other similar artifacts, in addition to limitations on incendiary weapons and blinding lasers; 1993 Convention on the Prohibition of the Development, Production, Storage and Use of Chemical Weapons; 2008 Convention on Cluster Munitions.

10. Melzer, *International Humanitarian Law*.

11. Melzer, 18–20.

12. Cinelli, *Direito Internacional Humanitário*.

13. International Committee of the Red Cross, "Como o Direito Internacional Humanitário define 'conflitos armados'?"

14. Rabkin and Yoo, *Striking Power*.

15. Ost, *O Tempo do Direito*.

16. ICRC, "A Guide to the Legal Review of New Weapons, Means and Methods of Warfare."

17. Ricupero, *A diplomacia na construção do Brasil*.

18. Ricupero.

19. The *uti possidetis* is a principle originating in Roman law. Portugal started using it with the objective of obtaining territorial gains in sharing and defining borders in South America, in confrontation with Spain. Subsequently, independent Brazil incorporated this strategy into its foreign policy.

20. Pinheiro, *Política externa brasileira*.

21. Brazil has been a temporary member of the Security Council for the biennia of 1946–47, 1951–52, 1954–55, 1963–64, 1967–68, 1988–89, 1993–94, 1998–99, 2004–5, 2010–11, and 2022–23. United Nations, "Brazil"; and United Nations, "Security Council Presidency."

22. United States of Brazil, Constitution of the United States of Brazil (February 24, 1891), § art. 88. According to this article, "The United States of Brazil, in no case, will engage in war of conquest, directly or indirectly, by itself or in alliance with another nation."

23. United States of Brazil, Constitution of the United States of Brazil (July 16, 1934), § art. 4. The article states that "Brazil will only declare war if the arbitration appeal does not fit or fails; and it will never engage in a war of conquest, directly or indirectly, for itself or in alliance with another nation."

24. United States of Brazil, Constitution of the United States of Brazil (September 18, 1946), § art. 4. The article states that "Brazil will only resort to war if it does not fit or fails to resort to arbitration or peaceful means of conflict resolution, regulated by an international security body, in which it participates; and in no case will it engage in a war of conquest, directly or indirectly, by itself or in alliance with another state."

25. Civil Office, "Federal Constitution," § art. 7.

26. Piovesan, "Artigo 4o, VIII, IX e X."

27. In 1996, Brazil published the first version of the *PND*. In 2005, a new *PND* was elaborated, followed by an *END* in 2007. In 2012, both documents were renewed. The new documents for 2020 remain under consideration by the National Congress as of late 2022.

28. Corte Interamericana de Direitos Humanos, Caso Gomes Lund e Outros ("Guerrilha do Araguaia") v. Brasil.

29. Ministry of Defense (MOD) and Joint Chiefs of Staff, MD34-M-03, *Manual de Emprego Do Direito Internacional Dos Conflitos Armados (DICA) Nas Forças Armadas*, 34–03.

30. Vianna and Camelo, "Defesa Cibernética no Brasil."

31. Markoff, "Disarmament and 'Peaceful Purposes' Provisions in the 1967 Outer Space Treaty," 3; and Vlasic, "Disarmament Decade, Outer Space and International Law."

32. Schmitt, "International Law and Military Operations in Space," 125; and Stephens, "The International Legal Implications of Military Space Operations: Examining the Interplay between International Humanitarian Law and the Outer Space Legal Regime," 80.

33. Schmitt, "International Law and Military Operations in Space."

34. The proposal aims to ban Space–Space and Space–Earth weapons.

35. Harrison, *International Perspectives on Space Weapons*.

36. Moltz, *Crowded Orbits*.

37. Melzer, *International Humanitarian Law*, 77.

38. MOD and Joint Chiefs of Staff, *Doutrina militar de defesa cibernética*.

Chapter 6

The Brazilian Air Force in UN Peace Operations
Ready to Support Global Stability

Pedro Henrique Nascimento dos Santos
Claudia Maria Sousa Antunes

Introduction

Peace operations typically are associated with troops on the ground acting with police and humanitarian workers from all over the world. Nevertheless, airpower plays an important role and can be considered an essential element for the development of the United Nations' peacekeeping and security activities. The Brazilian air force has participated in UN Peacekeeping Operations (PKO) since 1948, adopting a professional and determined posture among all its personnel to make humanitarian aviation a reality.

In the United Nations Peacekeeping Capability Readiness System (UNPCRS), contributing countries (such as Brazil) offer their capabilities and their level of readiness for the tactical, operational, or even strategic deployment of the means capable of applying those capabilities.[1] A particular predeployment training program is provided to national staff, airborne crews, and ground support personnel to improve PKO compliance with UNPCRS. This preparation establishes a condition of operational readiness and presents an opportunity to develop aerospace power strategic capabilities. This readiness also enhances the cooperative character of Brazil's international relations and the ability to project aerospace power.

The professional training of each component of an air unit considers the capabilities, employment attributes, missions, and air tasks to be fulfilled in the missions.[2] However, according to Dorn's analysis,[3] preparation for peace operations requires a protocol complementary to the technical and professional training developed by air forces. Human resources must be ready to act in unstable scenarios and with cultural diversity amid humanitarian crises typical of these operations. The specificity of the purpose demands doctrinal guidance based on practices and lessons learned in other peace operations.

Supporting Global Stability

Over the years, and through the presentation of a naturally pacifying stance, Brazil has obtained the ability to project itself beyond its own borders without inflicting damage or causing interference to the sovereignty of other nations. Brazil also projects itself by respecting the self-determination of peoples, by a policy of nonintervention, by recognizing equality among states, and by the peaceful resolution of conflicts and the defense of peace, all constituent parts of the constitution of the Brazilian Republic.[4] These principles are concomitantly aligned with the purposes of the United Nations organization established in Article 1 of the UN Charter:

The Purposes of the United Nations are:

1. To maintain international peace and security, and to that end: to take effective collective measures for the prevention and removal of threats to the peace, and for the suppression of acts of aggression or other breaches of the peace, and to bring about by peaceful means, and in conformity with the principles of justice and international law, adjustment or settlement of international disputes or situations which might lead to a breach of the peace.

2. To develop friendly relations among nations based on respect for the principle of equal rights and self-determination of peoples, and to take other appropriate measures to strengthen universal peace.

3. To achieve international cooperation in solving international problems of an economic, social, cultural, or humanitarian character, and in promoting and encouraging respect for human rights and for fundamental freedoms for all without distinction as to race, sex, language, or religion.

4. To be a center for harmonizing the actions of nations in the attainment of these common ends.[5]

The UN accomplishes this purpose by working to prevent conflict, helping parties in conflict make peace, deploying peacekeepers, and creating the conditions to allow peace to hold and flourish. These activities often overlap and should reinforce one another to be effective.

The Brazilian state combines its traditional cooperative reputation with national defense objectives that aim to contribute to regional stability and international peace and security.[6] The National Defense Strategy demonstrates that the armed forces' preparation and participation in peace operations contribute to regional stability and international insertion.[7] These dynamics receive special emphasis as strategic defense actions.

As pointed out by Cannabrava,[8] Brazilian engagement in peace operations also represents the fulfillment of international obligations. The practice of international cooperation activities characterizes Brazilian foreign policy as active and responsible in the search for global stability.

Aerospace Power and Operations Other than War

The term "peace operations" represents a concept that has evolved significantly in recent decades, as it presents an adaptive nature, which varies according to the needs of humanity and the political-strategic interests of states. The same can be said about aerospace power.

According to Boyne, aerospace power consists of an evolution of the concept of airpower,[9] since it deals with the use of space through satellites and intercontinental ballistic missiles, enabling the capacity to conduct military, commercial, or humanitarian operations. This concept is further defined by Chun as the exploration of the environment above the earth's surface by aerospace vehicles or devices to conduct operations in support of national objectives.[10] The Basic Doctrine of the Brazilian Air Force, described in the 1st Guideline of the Air Force Command, DCA 1-1,[11] does not differentiate the term airpower from aerospace power, instead defining them in a similar way: "Aerospace Power is the projection of National Power that results from the integration of the resources available to the Nation aiming at the use of airspace and outer space, either as an instrument of political and military action, or as a factor of economic and social development, to conquer and maintain national objectives."[12]

Brazil's participation in UN peacekeeping operations represents a relevant opportunity for it to project its aerospace power. It plays, in the contemporary scenario of relations between states, a decisive role that extends beyond its purely warlike application in war between nations and in regional or intranational conflicts.

The use of aerospace power in operations other than war (OOTW) has proven to be of relevant strategic and diplomatic gain in humanitarian and PKO. Hillen pointed out that there are applications along the spectrum of air activities in which there is the possibility, or even the unequivocal need, for the use of aerospace power.[13]

Within this spectrum of activities, from peace situations to those occurring in imminent war scenarios, PKO, also known as peace missions, present themselves as those that allow for the preparation and effective application of air force employment capabilities. This environment is described by Dorn as adverse and challenging, while sparing the excessive cost of national resources and the lives of pilots, crew members, and support teams under the judgment of enemy fire and the uncertainties of war.[14]

Brazil in PKO

Since 1945, Brazil has been actively present with diplomats, military, civilians, and police in peacekeeping missions. These missions develop under the aegis of the UN in favor of humanitarian objectives, in the resolution of disputes, and in actions for regional stabilization.[15] Brazil is among the 51 founding countries of the United Nations and currently remains as one of the 193 member states that compose it. Bracey, when examining the motivations for Brazil's contribution to peacekeeping operations, realizes that prestige and reputation as a regional leader, as well as increased projection of political power, are relevant factors in determining the participation of the country in such peace and security activities.[16]

Brazilian participation in peace operations establishes an advantageous relationship for the country. This occurs since the global cost is relatively low when compared to the positive factors of a political, strategic, and military nature. One of these advantages is the possibility of preparing and employing the armed forces in circumstances close, or even similar, to those found in a scenario of true belligerence. That provides a significant increase in the visibility of its structure and capacity to employ military means beyond its borders. Brazilian participation also can spur a political and economic extension of Brazilian influence in the international arena, especially as a peaceful agent that seeks a harmonious relationship and the mutual progress of nations. This perspective is according to openly declared objectives in documents of strategic dissemination such as the Defense White Paper and the Brazilian Federal Constitution itself: "Through dissuasion and cooperation, Brazil will thus strengthen the close link between its defense policy and its foreign policy, historically focused on the cause of peace, integration, and development."[17]

Brazil shares in its legal regulation concepts, principles, and values that are present in various aspects related to the UN PKO. According to Aguilar, "the decision to participate in a peace operation is motivated by factors ranging from humanitarian aid, devoid of any other intention, to the achievement of political objectives."[18]

Humanitarian-Biased Air Force

Brazil took part in at least 50 UN peacekeeping missions during the second half of the twentieth century. This has provided a remarkable increase in its logistical deployment capacity for highly humanitarian missions. This posture culminated in what might be termed the Brazilian way of peacekeeping,[19] a standard in which soft power (the ability to influence politically through ideological and cultural means) is used in a balanced way with hard power (the capacity for coercive action using economic and military resources) to achieve operational objectives related to the mission mandate.[20]

Brazilian troops, in the form of platoons, companies, and battalions, have participated in UN missions in Egypt, Mozambique, El Salvador, Angola, East Timor, and Haiti.[21] This participation generates a progressive gain in the capacity of the Brazilian state's means of defense in terms of modernization, professionalism, training, force projection, and humanism. Brazil's navy and army, in particular, perceive multifactorial gains from their participation in United Nations PKO.

The UN Naval Task Force in Lebanon has been commanded by Brazil since its installation in 2010, with naval resources and crews prepared by the Brazilian navy. This participation has exponentially increased naval logistical projection capacity, competence in leading joint operations, and improved employment techniques of its personnel in international waters. In the same way, the Brazilian army has developed capabilities due to the need to prepare and employ troops for peace missions in Suez, Egypt (United Nations Emergency Force 1, UNEF I, 1957–67); Mozambique (United Nations Operation in Mozambique, UNMOZ, 1992–94); Angola (United Nations Angola Verification Mission 3, UNAVEM III, 1995–97); and Haiti (Mission des Nations Unies pour la Stabilization en Haiti, MINUSTAH, 2004–17).

The Brazilian air force's human and material support to the UN has been in place since the event marking the start of UN Peace Operations: the United Nations Special Committee on the Balkans (UNSCOB), in 1948. This mission included the participation of FAB Captain João Camarão Telles Ribeiro and two other military officers from the army and navy.[22] This group of international observers would monitor any violations of the signed peace agreement in that region. From then on, the FAB has participated in a relevant manner in PKO under the aegis of the United Nations and multilateral organizations.

This involvement occurred with specially trained troops assigned to UN Operational Control to act in each of the peace operations.[23] The army, navy, air force, and auxiliary forces (such as military police and state fire departments) have participated in several other UN peace missions, contributing observers, staff components, and specialists (as with Ribeiro and his team with UNSCOB). Continuing this longstanding contribution, over the last decade Brazil's participation in PKO has helped further develop the professional and specialized character of these forces.[24]

The history of FAB participation in peace operations is extensive and successful. However, these actions were mostly in logistical support, except for the crews and teams of H-19 helicopters in the Congo (Organização das Nações Unidas no Congo, ONUC) during the 1960s, which carried out medical and casualty evacuation missions, alert, reconnaissance, and liaison under the operational control of the UN. (This occasion gave rise to the FAB's celebration of Rotorcraft Aviation Day on February 3, 1964.)[25]

Peacekeeping Readiness

Before being engaged in peacekeeping operations, troops and units must go through a specific process allowing appropriate preparation for modern peacekeeping missions. The UN Capabilities Provision and Resource Readiness System (UNPCRS) was conceived in 2015 as part of a broad institutional reform aimed at improving the structure and execution of UN Peace Operations (table 6.1). This system replaced the United Nations Standby Arrangement System (UNSAS).[26]

Table 6.1. UN Peacekeeping Capability Readiness System

Level	Criteria
1	The Troop Contributing Countries (TCC) make a formal pledge to a unit along with (1) the table of organization, (2) a list of major self-support equipment, and (3) details of specialists (if any) certification of completion of basic training. Member states may include police and any nonmilitary capabilities. Emerging TCC and aspirants who do not yet meet these basic requirements will not be registered at Level 1 but will be engaged as part of a preparatory process managed by the Force Generation and Capability Strategic Planning Cell.
2	Based on UN operational requirements, Level 1 pledges will be elevated to Level 2 through a preliminary memorandum of understanding (MOU) negotiation process and an assessment and advisory visit (AAV) by a UN headquarters team composed of members of the Force Generation Service and the Operational Support Department.
3	After a satisfactory AAV, only units that have achieved a reasonable degree of readiness will be upgraded to Level 3. A detailed MOU will be negotiated, and the TCC will provide a load list as required by the Department of Operational Support for planning.
Rapid deployment	Upon reaching level 3, the TCC can promise to deploy the mission within 30/60/90 days of the request made by the UN headquarters. In the case of facilitators, this level will also be linked to the offer for rapid deployment (if they are declared eligible for a specific mission according to the offer guidelines).

Adapted from United Nations. "Guidelines: Peacekeeping Capability Readiness System (PCRS)." United Nations Department of Peace Operations, Department of Operational Support, 2019. https://pcrs.un.org/.

Brazil is among the countries able to contribute troops to UN Peace Operations; these countries are known as TCC (Troop Contributing Countries) and, as part of UNPCRS, Brazil is capable of furnishing its own air unit. However, although air operations are fundamental to the functioning of peace missions from the tactical to the strategic level,[27] Brazil does not currently have air units employed in peace operations under the operational control of the UN. Other countries such as South Africa, Argentina, Canada, Honduras, Uruguay, Chile, Sri Lanka, India, Ghana, and Rwanda have done so or do so with fixed-wing and rotary-wing units or even with remotely piloted aircraft.[28]

Brazilian aerospace power has been used sporadically in peace operations without the transfer of operational control to the UN (see table 6.2). The most notable situations occurred in Congo in 1960 (UNTEF II) and 1993 (Operation Artemis), East Timor in 1999 (INTERFET) and 2002 (UNMISET), and Haiti from 2004 to 2017 (MI-

NUSTAH). The latter operation made evident the need for a strategic cargo aircraft for the deployment of troops over long distances,[29] and Operation Artemis, despite being approved by the UN Security Council, is not classified as a UN peace operation, since it was conceived and led by the European Union.

Table 6.2. Brazilian participation with aircraft in UN peace operations

Period	Location	Peace operation	Air means
1957–1967	Egypt	UNEF-I	B-17G, C-54G
1960–1964	Congo	ONUC	C-47, H-19
1993–1994	Mozambique	UNOMOZ	C-130
1995–1997	Angola	UNAVEM III	KC-137, C-130
1999	Congo & Uganda	Artemis[30]	C-130
1999–2005	East Timor	UNTAET / UNMISET	KC-137
2004–2017	Haiti	MINUSTAH	KC-137, C-130, C-99

Adapted from Gonçalves, "A Força Aérea Brasileira na Missão de Estabilização das Nações Unidas no Haiti: A Dependência de uma Aeronave de Transporte Estratégica."

The motivations that lead countries to contribute air units to PKO are varied. Among them are the search for regional political influence, economic advantage from the reimbursement of the units employed, geostrategic factors related to the projection of force, and humanitarian issues of international assistance. Between 2000 and 2016, air assets in UN PKO proved to be fundamental for conducting missions and implementing a wide range of mandates.[31]

Despite the great importance of the material and technological factors inherent in a capability offer for the UNPCRS system, the fundamental element to which the readiness protocol applies is the human factor. In 2018 and 2019, Brazil adopted its readiness protocol to keep human resources available in the UN capability readiness system. In addition, the country maintains a medium utility helicopter air unit in constant readiness for potential UN deployment.

Aerospace power readiness results from the capability of constituent elements. A helicopter air unit, despite its great importance, represents only one element. An air unit has, in its basic formation, all the human and material resources geared towards the fulfillment of tasks that are already in the interest of the air force and national defense. This preparation can be divided in two main aspects: personnel training and material endowment. Continued training and doctrinal orientation for peace missions is one way for Brazil to increase its national aerospace power.

The provision of material refers to the aircraft, equipment, and supplies needed for the type of operation desired and depends on the acquisition process (the acquisition process refers to the procurement processes carried out by the public administration) and the specificity of the mission for which a given unit is destined. These characteristics remain unchanged from the beginning to the end of the operation.

Personnel training corresponds to the process to which pilots, crew members, support, maintenance, and security teams must be submitted to perform their duties well during a peace mission. Crew members—such as observers, special equipment operators, search and rescue teams, and flight mechanics—are important in the context of a peace mission. However, the maintenance and security teams maintain their routine activities in a manner like that usually adopted when on national territory.

Personnel are subjected to a significant impact on their routines when in a peacekeeping operation. This occurs for several reasons, such as the fulfillment of mission guidelines presented in documents such as status of forces agreements; mandates (i.e., operating guidelines and general objectives of the peace mission for all entities involved); status of unit requirements; Letter of Assist; rules of engagement; and memorandums of understanding.[32] Also, it results due to the interaction with international organizations and armed forces from other countries on missions and, mainly, due to the minimum requirements for pilot qualification demanded by the UN to act in peacekeeping missions.

Thus, preparing air units for UN PKO consists of doctrinal training for pilots and elements of the unit general staff. This procedure is preconized by the manual for military air units in peacekeeping operations.[33] This preparation also respects the precepts of the *National Defense Policy*: "Brazil, due to its tradition as a defender of dialogue and harmonious coexistence among peoples, will continue to be invited to make its contribution to world peace. As a result, it must be prepared to meet possible demands of participation in Peace Operations, under the aegis of the United Nations—UN—or multilateral organizations. This involvement, observing the circumstances of the moment, must follow the principles and priorities of Brazil's foreign and defense policies."[34]

Readiness and Strategic Management

The FAB's air units often operate in conditions close to those observed in a PKO environment. These units have both human and material resources fully prepared for the operational employment of the capabilities of a medium utility helicopter unit. To establish a minimum level of planning, it is necessary to consider some facets about strategic management aimed at preparing air units to act in UN PKO. The first refers to Brazil's situation. The country is among the five largest countries in land area and among the most populous. Its military force is considered the ninth most powerful in the world and, from this perspective, is the most militarily prepared country in Latin America.[35] Brazil has been present in UN Peace Operations since its creation. However, it operated with an air unit under UN operational control only between 1960 and 1964.

The second facet concerns the direction the country intends to take regarding participation in peacekeeping missions. According to constitutional precepts and the strategic concept of the air force 100,[36] the FAB must have the capacity to act in an international context through participation in peacekeeping operations. This participation must aim at contributing to world peace and to the fulfillment of international commitments. It can be conducted jointly with other nations, for international peace operations, through the consent of the parties in dispute, to reach a peace agreement, or to supervise the implementation of the terms of such agreements.

The third facet is what adjustments might need to be made to prepare teams for international peace operations. FAB air units have a high degree of professionalism and operational preparedness, but the question remains whether there is sufficient material resource endowment for the employment of aircraft. Adapting to act according to UN-specific guidelines would require doctrinal training of pilots and members of the unit staff. Brazil has a Joint Center for Peace Operations (Centro Conjunto de Operações de Paz do Brasil, CCOPAB), whose main purpose is to conduct doctrinal training of military personnel to act in the PKO environment. The CCOPAB is linked to the Ministry of Defense and recognized by the United Nations to carry out instruction and training aimed at the preparation of human resources, in accordance with the development of capabilities related to the most recent doctrinal guidelines presented by the UN.

The emphasis of the doctrinal training process is on manual analysis, lectures with members of the UN peacekeeping generation system, planning exercises, and evaluations. Measuring the results of this preparation is done by means of tactical exercises in simulated PKO scenarios. In addition, theoretical assessments are carried out at the end of the training process, with the consequent certification by CCOPAB. The preparation can also be measured by the performance of the air units during peace operations.

The preparation protocol for Brazilian military air units is carried out jointly by CCOPAB and the Air Force Preparatory Command (Comando de Preparo, COMPREP). CCOPAB presents doctrinal concepts and specific knowledge about the actions of military units involved in peace operations. This activity is carried out through the air units' preparedness instruction offered to the UNPCRS system (Instruções de Preparação de Unidades Aéreas, IPUNAER). COMPREP promotes the actions necessary for operational training and maintenance aimed at the effective employment of national air power capabilities for specific purposes.[37] The Operational Exercise Tápio provides training for conditions like those in UN peacekeeping missions.[38]

IPUNAER and Tápio exercises represent the preparation protocol activities adopted by Brazil to train human resources offered to the United Nations according to the UNPCRS system. The analysis of the stages and activities of the preparation aimed at the air unit components demonstrates that UN requirements are achieved.[39] The preparation protocol, which precedes the possible deployment of an aerial unit to PKO, favors the development of the aerospace power projection capacity and consolidates the traditional posture of the Brazilian state focused on international cooperation.

Thus, a condition of operational readiness is established. This readiness mutually favors Brazil's cooperative character in international relations and the exploitation of the dissuasion principle through aerospace power projection capacity. It is characterized, therefore, as a strategic action of national defense.

Conclusion

Aerospace power refers to a comprehensive concept that involves the capacity to exploit the airspace and space environment to achieve national goals. This capability is achieved through several components. This chapter outlined aspects of the strategic management of aerospace power, showing that the preparation of human and material resources for a PKO represents a way to act positively in pursuit of the fulfillment of national goals.

The history of the Brazilian presence in UN peace operations is long and successful. Preparing and deploying FAB air assets and human resources under the rules of the current UNPCRS system bring great strategic geopolitical opportunities for Brazil. However, this requires doctrinal training. The doctrinal training offered by the Brazilian Joint Peace Operations Center is the most effective way to meet the parameters required by the UN. The Air Units Preparedness Instruction offered to UNPCRS System (IPUNAER) and Operational Exercise Tápio effectively constitute the preparedness protocol established by the Brazilian air force to meet the UN's PKO requirements.

The Brazilian constitution establishes principles of international relations associated with the self-determination of peoples, nonintervention, and defense of peace. Due to these factors, the preparation of a military air unit to act in peace operations is one of the most accessible ways to qualify human and material resources to act in conditions close to a real belligerent scenario.

Notes

1. United Nations, "Guidelines: Peacekeeping Capability Readiness System (PCRS)."
2. Air Operations General Command, "DCAR 100B Capacitação de Recursos Humanos No Âmbito Do COMGAR."
3. Dorn, *Air Power in UN Operations*.
4. Lopes, "Breves Considerações sobre os Princípios Constitucionais das Relações Internacionais."
5. United Nations, "Charter of the United Nations and Statute of the International Court of Justice," chap. 1, Purposes and Principles.
6. Ministry of Defense (MOD), *Livro Branco de Defesa Nacional*.
7. MOD.
8. Cannabrava, "O Brasil e as operações de manutenção de paz."
9. Boyne, *The Influence of Air Power Upon History*, 18.
10. Chun, *Aerospace Power in the Twenty-First Century*.
11. MOD, DCA 1-1.
12. MOD, 35.
13. Hillen, "Peacekeeping at the Speed of Sound."
14. Dorn, *Air Power in UN Operations*, 232.
15. Aguilar, "A Participação Do Brasil Nas Operações de Paz."
16. Bracey, "O Brasil e as operações de manutenção da paz da ONU."
17. MOD, *Livro Branco de Defesa Nacional*, 51. This is the most complete document on Brazil's defense activities. It aims to clarify for Brazilian society and the in-

ternational community the policies and actions that guide security procedures and protection of sovereignty.

18. Aguilar, "A Participação Do Brasil Nas Operações de Paz," 124.
19. Kenkel, "Brazil's Peacekeeping and Peacebuilding Policies in Africa," 275.
20. Nye, "The Information Revolution and Power."
21. A platoon is the smallest military unit under command of an officer. It is typically composed of one noncommissioned officer, one staff sergeant, and 20 to 50 soldiers. A company is equivalent to two to eight platoons commanded by a major or captain, and a battalion is equivalent to two to six companies, commanded by a lieutenant colonel or major.
22. Bittencourt, "A presença da Marinha do Brasil em missão pioneira de manutenção de paz."
23. Operational control is the segment that directly designates mission accomplishment.
24. Braga, "MINUSTAH and the Security Environment in Haiti," 715.
25. Aerospace Museum, "Dia da aviação de asas rotativas."
26. United Nations, "Guidelines: Peacekeeping Capability Readiness System (PCRS)."
27. Dorn, *Air Power in UN Operations*.
28. Novosseloff, *Keeping Peace from Above*.
29. Gonçalves, "A Força Aérea Brasileira na Missão de Estabilização das Nações Unidas no Haiti," 111.
30. Operation Artemis is considered a peace operation due to its peaceful character and because it was approved by the UN Security Council in 1998. However, it was fully coordinated by the European Union because it involved the direct interest of the European Union in stabilizing the humanitarian crisis in the Congo.
31. Novosseloff, *Keeping Peace from Above*.
32. Letters of Assist refers to a contractual document issued by the United Nations to a government authorizing it to provide goods or services to a peacekeeping or other UN operation.
33. United Nations, *United Nations Peacekeeping Missions Military Aviation Unit Manual*.
34. Brazilian National Congress, *National Defense Policy and National Defense Strategy, 2020*, 32.
35. Global Fire Power, "2021 Military Strength Ranking."
36. MOD and Air Force Command, "Concepção Estratégica."
37. Air Operations General Command, "DCAR 100B Capacitação de Recursos Humanos No Âmbito Do COMGAR."
38. Air Force Command, "Relatório Final de Exercício: ExOp Tápio II."
39. United Nations, *United Nations Peacekeeping Missions Military Aviation Unit Manual*.

Chapter 7

Interoperability among Brazil's Armed Forces
The Cultural Intelligence Perspective

Marta Maria Telles
Alessandra Veríssimo Lima Santos
Rainer Ferraz Passos

Introduction

This chapter examines the interoperability among Brazil's armed forces through the lens of cultural intelligence. Recent national and international studies have discussed the scenario of armed forces (AF) in joint operations considering three points of view: their process of jointness, understood as a collaboration among forces to improve operational effectiveness,[1] the concept of interoperability,[2] and the influence of organizational culture in the military environment.[3]

The theoretical assumptions embedded in the framework of those cited studies enabled the formulation of ideas for a conceptual approach, proposing the process of jointness as the link between cultural intelligence and interoperability, with the expectation of fostering reflection on the possible convergences and contributions of cultural competencies for efficiency in joint actions.

The Brazilian context concerning joint employment in training and humanitarian aid missions and/or peace operations demands, in addition to doctrinal knowledge, assimilation and adaptation to other cultures as well as social interaction, since current scenarios require flexibility and versatility.[4] By 2008, the head of joint operations at the Ministry of Defense (MOD) pointed out that "the different organizational cultures and the long period of sparse integration between AF are obstacles to the evolution of joint/combined doctrinal concepts."[5]

In that regard, the *National Defense Strategy (Estratégia Nacional de Defesa, END)* highlights among the competencies required of combatants the knowledge, skills, and attitudes that support integrated action and adaptation to the peculiarities of other forces to ensure doctrinal unity with the combination of means and the convergence of efforts.[6]

The Joint Staff of Brazilian Armed Forces (Estado-Maior Conjunto das Forças Armadas, EMCFA), created in 2010 as the military division of the MOD, is responsible for joint employment planning. In 2010, the term "joint" was added to the official nomenclature.[7] Over the years, EMCFA has sought to adjust its structure so that it is sufficiently able to advance effectiveness of the joint employment, according to the currently assigned scenario.[8]

To this end, EMCFA intensified training operations, for example Operational Exercise Tápio, held in Brazil in 2019 and 2020, which focused on irregular warfare missions like those of United Nations (UN) peacekeeping missions, to ensure the maintenance of qualifications and operational training.[9]

Among the international exercises, 2019's Green Flag West, whose goal was to simulate combat tactics against threats in regular and irregular war, allowed the Brazilian military to learn more about the US Air Force and its way of operating in those scenarios.[10] The Culminating Exercise, completed in February 2021 and developed through a five-years-long common joint plan of activities between Brazil and the United States, aimed to prepare military personnel and crew members for airborne operations.[11]

It is undeniable that training in tactics, techniques, and procedures is vital. However, people represent a critical factor in operational contexts, as in any other context. The actors involved in operational theaters need to rely on each other's service capabilities, as this is the essence of interoperability. This reliance is a vital part of deepening confidence and represents the technical side of "trusting." Proficiency in service capabilities is a *sine qua non* condition embedded in the process of jointness, because otherwise it would be difficult to build trust and understanding. Neither are acquired through a doctrine but rather derive from bonds created whether in training or in real combat.[12]

In joint operations, building bonds, trust, and understanding comes from contact with the individuals and cultural contexts of each force. These factors inform the concept of cultural intelligence as it encompasses both intelligence and emotional quotients, which consist of knowledge and skills acquired about a given culture for more effective interaction.[13]

Therefore, when analyzing the management of joint operations, it is important to look at the human component and its relationship with the cultural component, because the success of organizational activities is increasingly based on the ability to effectively deal with intercultural interactions.[14] Interculturality presupposes communication between differences. This should mean that the counterpart is understood from a cultural point of view, beyond the solely linguistical or procedural, and that a real dialogue is established, whether with words or actions.

Based on this understanding, cultural intelligence becomes fundamental in contemporary military activity in two different ways. First, superiority over the opponent nowadays may be achieved in a way other than the traditional combat. "The impact is not limited to the war action itself, but it also takes place with the cultural aspect, noticeable at the phases of approach, invasion and occupation, until pacification, interacting with societies."[15] Next, being culturally intelligent is fundamental because synergy among jointly operating AF—along with continuous preparation, suitable means, and prompt response capability—is the way to obtain the maximum performance of aerospace, land, and naval powers.[16]

In this sense, "aerospace power can play a major role in helping the joint force achieve its objectives with less risk in many scenarios across this operational continuum."[17] Gathering airspace information allows for the maintenance of situational awareness so that armies and navies can better shape their own actions as well as foresee or respond to the actions of the adversary. Overhead collection systems, whether space-based or manned and unmanned aerial platforms, can generate data from sources within a joint operational context, useful for planning, decision making, and action-directing.[18]

Jointness, Interoperability, and Cultural Intelligence: An Approach

According to the literature, interoperability is understood as the competence used by individuals and organizations to operate together in a coherent, firm, and efficient manner to achieve common objectives, whether tactical, operational, and/or strategic.[19] Its development "seeks to optimize the use of human and material resources, as well as to improve the employment doctrine of the Armed Forces."[20]

In turn, jointness strives for concentrating all these efforts, whether in environments of peace, crisis, or war, to increase the effectiveness of military operations.[21] Jointness depends on people and their ability to perform the actions required as service capabilities.[22]

Although the concepts of cultural intelligence and interoperability are distinct, their foundation on culture makes them similar. Interoperability not only encompasses the ability of militaries to work together even though they are culturally different, but also requires understanding a common organizational environment.[23]

This is a challenge because each AF has its own culture, with distinct views on conflict scenarios and principles and scale of values that direct their actions in contexts of peace or war.[24] As Lemos writes, "The individual cultures of each AF, their *ethos*, remain strong and pursue the ideal accommodation, catalyst for a perfect synergistic integration."[25]

The concern with this synergistic integration is reflected in the annual development of activities as part of the programming of joint events of the MD, involving cadets and midshipmen, at Brazilian military academies, up to colonels and navy captains.[26]

Following from this focus of interoperability encompassing the understanding or construction of a common organizational environment, EMCFA efforts in intensifying joint training are crucial for strengthening jointness and interoperability as well as building up an environment conducive to the development of cultural intelligence skills.

After operations in Afghanistan and Iraq, the US Navy noted that commanders and soldiers lacked sufficient cultural skills, demonstrated by the high number of lives lost and the numerous resistance actions against the presence of the US forces in occupied territory. As a result, the US Department of Defense created training centers and cultural intelligence development programs for cadets and officers, conducted by social scientists and military instructors. The objective was to enable them to analyze the area of activity, whether in terms of people, culture, or economic development.[27]

This example indicates that crossing a cultural boundary requires a learning process, whether formal or informal, systematic or unsystematic.[28] Thus, among the actions that contribute to culturally intelligent behavior, training, study, observation, and reflection are critical.[29] This adaptive ability allows the understanding of each interlocutors' culture and the interpretation of ideas, acts, and intentions, allowing individuals to adapt their behavior to the different perspectives they have experienced.[30]

The after-action review (Análise Pós-Ação, APA) of Operational Exercise Tápio—which, since 2019, has involved joint activities among the Brazilian armed forces—indicated improvement in team preparation and the self-confidence of individual military members in performing their tasks for the fulfillment of missions. In the words of the exercise director, "there was an evolution of maturity, both of knowledge and skills, especially of the attitudes of the crew members who were in a dynamic and challenging scenario."[31] From a behavioral learning perspective, the demand for a process to learn how to cross cultural boundaries is also perceived here, as indicated in the example of operations in Iraq and Afghanistan.

The resulting impressions from the evaluation of Exercise Tápio enable some parallels to be drawn. The first concerns jointness and interoperability, given that exercises collaborate to mitigate the failure to which missions are subject, whether by the human or technological component, and to seek the effectiveness of operations. Synergy should reflect joint thinking, jointness, defined by Lemos as "power capable of integrating the strengths of the capabilities of at least two Forces, which make up AF of a State, through coordinated effort to achieve a common goal."[32]

Considering cognitive learning, Marquis, Dye, and Kinkead recommend that the military, at all levels, study and recognize the benefits and challenges of joint capacity.[33] In the Brazilian scenario, Pires understands the need for academic systematization and suggests "the establishment of a joint curriculum."[34] A joint curriculum should gather technical knowledge about weapons systems of each armed force; support in the theater of operations; gains derived from the joint use of such systems; effects and implications of information and cyber operations at the tactical level; and doctrinal knowledge, both individual and joint.

The second parallel aims to bring the cultural component closer to the forefront of military training, since "becoming culturally intelligent is essentially learning by doing."[35] Earley and Mosakowski explain cultural intelligence from the cognitive, emotional-motivational, and physical dimensions. According to the authors, cultural intelligence resides in head, body, and heart.[36]

The "head" corresponds to the knowledge that the individual holds and the knowledge that one can obtain. The "heart" represents the self-confidence and motivation that direct one to achieve success and is anchored in the domain of a particular task or a set of circumstances, while the "body" transforms one's intentions into behaviors or actions.[37]

These three elements can be identified in the APA performed at the end of the Tápio exercise each year. The heart is made explicit by trust in oneself and in others. The head comprises the technical mastering achieved during the exercise, with the increased level of knowledge and skills. The body represents the way the military acts and reacts when facing the dynamic and challenging scenarios within the exercise.

The two notes indicate the relevance of the dynamic between human and cultural components. In such integration scenarios, because of each cultural context of the armed forces and the social contexts for their action, the exercise encourages more careful reflections on ways to better explore the applicability of cultural intelligence.

This dynamic can also be illustrated by a study on the adoption of cultural intelligence in the defense of the Brazilian area of the Amazon forest. Considering that the success of military operations in the northern region of the country depends on the support of the population, mainly represented by the indigenous movement, the study pointed out that the connection between the armed forces and the movement is hampered by mutual misunderstanding, prejudice, and interference of social factors.[38] Establishing a "permanent dialogue,"[39] signaling that the interaction between the indigenous and military populations required something besides the role of interpreters, verbal fluency, or technical knowledge in the local language, meant investigating social differences and their consequences on the behavior and expression of the two groups.[40] The relevant political profile of the indigenous people leads to the need for cultural training of soldiers with regard to "indigenism, based on social sciences such as anthropology and sociology, and by strict observance of the legal provisions in force."[41]

The theoretical knowledge about culture and indigenous beliefs strengthens proficiency in service capabilities and in the list of cultural intelligence skills, which is understood as "an instrument that leads to the proper development of relations, by favoring conflict resolutions and knowledge of individual and collective identities."[42]

Trust and understanding, derived from service competence,[43] are fundamental bases for the perception of local reality and for the creation of bonds with the indigenous. Endowed with the capacity for argumentation and persuasion, they expect the same willingness to dialogue by military leaders.[44] In their actions, "head," "heart," and "body" should be the focus in the search for a posture attentive to these interactions. The singularity of the practice of cultural intelligence lies in the possibility of consciously recognizing cultural differences.[45]

Another example is the Brazilian participation in Operation Artemis, from July to September 2003. The European Union conducted this operation in the pacification process in the Democratic Republic of the Congo, with the objective of providing support to regional peace-making efforts. Two C-130 aircraft were sent with 41 military personnel on board.[46]

According to a report by former members of the mission, to define the criteria for troop selection, military authorities considered the intelligence reports upon the geopolitical scenario of an escalating humanitarian crisis, with an increasing number of refugees and conflicts between ethnic groups. Recognizing these cultural differences, they concentrated their efforts on identifying military personnel with skills that would allow them to act safely in these adverse conditions and help achieve the success of the mission.[47]

Criteria were established for the selection of the military, and the concern with service skills regarding knowledge about C-130 Hercules aircraft and the ability to operate in harsh locations was observed, considering the complexity of the operational scope. It is also understood that the leadership considered cultural skills when selecting more experienced military for the mission in dealing with foreign troops. Although Operation Artemis has not been carried out recently, reports of former members highlighted the importance of the crew experience in similar situations, the possibility of applying knowledge in a real mission, and the relevance of training. Those observations demonstrate that technical competence combined with cultural skill is relevant in the context of military operations.[48]

Thus, the elaboration of joint thinking as a system, focused on implementing the idea of interoperability within different cultural contexts, suggests the insertion of new practices, for example, cultural intelligence–related mission tasks.

Conclusion

The proposed concepts elaborated in this chapter aimed to deepen the discussion on the importance of considering, in addition to the doctrinal, the human and cultural components in joint operations. Taking interoperability as a starting point and the process of jointness as a guiding thread, mutual trust was illustrated to be a central and essential aspect of convergence of these concepts and cultural intelligence.

It is noteworthy that a limitation of this reflective approach is the impossibility of disclosing data from official reports of training and joint operations, since they are classified. Thus, only data available from public domain sources were used to contextualize the discussions.

Nevertheless, bringing to discussion and into light the idea of self-confidence and mutual trust as basic aspects to be improved and measured as additional indicators for improving joint operations and consequent efficiency in conflicts may offer new ideas for courses, exercises, or operational planners to define objectives and measures of efficiency. Moreover, this elaboration may draw attention to behaviors or beliefs regarding confidence and trust that might have been overlooked previously.

Therefore, it is expected that the continuation of this discussion would evolve into observations of these aspects in courses and operations, as well as to papers about experiences with formal inclusion of practical objectives that could improve self-confidence and mutual trust in joint operations, exercises, and courses.

At most, it is expected that this chapter will contribute to discussions of deepening the conceptual framework on the applicability of cultural intelligence in contexts of joint operations, furthering understanding on how to take it best in such a complex operational environment.

Notes

1. Lemos, "Jointness, pensamento conjunto e conjuntez"; Freire and Baccarini, "Uma análise neo-Institucional da adoção de jointness pelos Estados Unidos"; Marquis, Dye, and Kinkead, "The Advent of Jointness During the Gulf War"; Pessoa and Freitas, "A Adoção Do Modelo Joint"; and Vitale, "Jointness by Design, Not Accident."

2. Teixeira and Freire, "A importância da Interoperabilidade como Instrumento de Convergência nas Operações Militares do Brasil"; Costa, "Interoperabilidade: o impacto da diminuição das ações com tropas e meios nas operações de adestramento conjunto"; and Ciocan, "Perspectives on Interoperability Integration within NATO Defense Planning Process."

3. Wortmeyer, "Introdução a Uma Perspectiva Psicossociológica Para o Estudo Das Forças Armadas"; and Oliveira, "Ensino e interoperabilidade."

4. Ministry of Defense (MOD), DCA 1-1.

5. Costa, "Interoperabilidade"; and Baptista, "Ministério da Defesa: Realidade, Desafios e Perspectivas," 53. According to Baptista, the different organizational cultures and the long period of sparse integration between the armed forces are obstacles to the evolution of joint/combined doctrinal concepts.

6. Air Force Chief of Staff, "Exercício simula cenários de guerra e treina militares para situações reais."
7. Lemos, "Jointness, pensamento conjunto e conjuntez," 2.
8. Santos, "Prontos a Qualquer Hora."
9. Santos, 25.
10. Iha, "FAB encerra participação no Exercício Internacional Green Flag West."
11. Vasconcellos, "Com balanço positivo, KC-390 conclui participação no Exercício Culminating."
12. Wilkerson, "What Exactly Is Jointness?"
13. Thomas and Inkson, *Inteligência Cultural - Instrumentos para Negócios Globais*; and Thomas et al., "Cultural Intelligence: A Theory-Based, Short Form Measure."
14. Thomas et al., "Cultural Intelligence," 3.
15. Calaza, "Inteligência cultural: novos parâmetros na formação do oficial ante a nova geração de conflitos" 48. According to the author, the impact is not limited to the war action itself but also takes place with the cultural aspect, noticeable at the phases of approach, invasion, and occupation, until pacification, interacting with societies.
16. MOD, DCA 1-1.
17. Schwartz and Stephan, "Don't Go Downtown without Us," 3. According to the authors, "aerospace power can play a major role in helping the joint force achieve its objectives with less risk in many scenarios across this operational continuum."
18. Schwartz and Stephan, 6.
19. Ciocan, "Perspectives on Interoperability Integration within NATO Defense Planning Process," 4; Teixeira and Freire, "A importância da Interoperabilidade como Instrumento de Convergência nas Operações Militares do Brasil," 4; Oliveira, "Ensino e interoperabilidade," 4; and Lemos, "Jointness, pensamento conjunto e conjuntez," 24.
20. MOD, "Glossário das Forças Armadas MD35-G-01."
21. Vitale, "Jointness by Design, Not Accident," 214; and Marquis, Dye, and Kinkead, "The Advent of Jointness During the Gulf War," 4.
22. Wilkerson, "What Exactly Is Jointness?," 4.
23. Ispas and Tudorache, "Cultural Interoperability."
24. Builder, *The Masks of War*.
25. Lemos, "Jointness, pensamento conjunto e conjuntez," 22.
26. Pires, "A importância da Reforma 'Goldwater-Nichols,'" 5.
27. Depaula, "Predictores Globales De La Performance De Estudiantes Militares."
28. Wortmeyer, "Introdução a Uma Perspectiva Psicossociológica Para o Estudo Das Forças Armadas," 5.
29. Thomas and Inkson, *Inteligência Cultural*, 5.
30. Lopez, *Negociação e a Inteligência Cultural*.
31. Air Force Agency, "EXOP Tápio totaliza cerca de 1.200 horas de voo." According to this news release, there was an evolution of maturity, both of knowledge and skills, especially of the attitudes of the crew members who were in a dynamic and challenging scenario.
32. Marquis, Dye, and Kinkead, "The Advent of Jointness During the Gulf War," 6.
33. Pires, "A importância da Reforma 'Goldwater-Nichols,'" 40.
34. Thomas and Inkson, *Inteligência Cultural*, 9.
35. Thomas and Inkson, 9.
36. Earley and Mosakowski, "Cultural Intelligence."
37. Earley and Mosakowski.
38. Visacro, "Inteligência cultural."
39. Visacro, 8.
40. Costa, "Interoperabilidade," 8.

41. Matos and Matos, "Grupos Indígenas y Militares."
42. Visacro, "Inteligência cultural." According to the author, indigenism is based on social sciences such as anthropology and sociology and by strict observance of the legal provisions in force.
43. Santos, "Inteligência cultural," 133.
44. Wilkerson, "What Exactly Is Jointness?," 8.
45. Thomas and Inkson, *Inteligência Cultural*, 8.
46. Santos and Jayme, "Operação Artemis."
47. Santos and Jayme, 61–62.
48. Santos and Jayme, 61–62.

Chapter 8

Geopolitics, Culture, and Law
Epistemological Convergences that Impact Aerospace Power

Guilherme Sandoval Góes
Maria Célia Barbosa Reis da Silva

Introduction

This chapter deals with the epistemological dialogues about geopolitics, culture, and law in contemporaneous times and how they interact, not only with aerospace power but with state power as well. Epistemology, the study of knowledge, provides a framework for studying the trends of geopolitics, culture, and law, tracing their intersections that affect a state's national power. This kind of study is important for a country like Brazil and for its aerospace power, represented here by civil aviation, aerospace infrastructure, the aerospace scientific-technological complex, the Brazilian air force (Força Aérea Brasileira, FAB), space program and aerospace industry, among others.

Culture permeates all knowledge; it is the identity record of any macro or micro group. Every human behavior fits or has its mark on culture: politics, economics, religion, food, arts, fashion, legal codes; individual and collective bearings and values keep cultural traces. Culture is porous, as it allows fluxes of exchanging information, while different cultures when in contact may influence and be influenced by others, as epistemological flows in a circular and reciprocal exchange of information, beliefs, and behaviors.

Human beings adapt to the individuals who are part of their collective and to their environment, both of which in turn define their identity and interactions within their collective and with other social groups. Such interactions make a rubric of themselves and the group to which they belong. The pattern that emerges from all those cultural mixes and combinations is variable over time, representing a zeitgeist, the spirit of the period for certain populations in certain geographic regions.

One must be attentive to note the cultural mixes so peculiar to postmodernity: what remains of each one, what changes, and the cultural combinations that emerge from them to proliferate in other styles more in tune with the Zeitgeist. In other words, the scientist must consider the cultural patterns of certain people at specific time periods.

Geopolitics is the study of how the planet's geography influences countries' political relations internally and internationally. It explains much of the complexities of international relations and balance of power.[1] To determine the linkages of state power, including its aerospace power, this article examines national culture, legal-constitutional order, and multidisciplinary approaches that systematize reciprocal influences between geopolitics and culture (geoculture) and between geopolitics and law (geolaw).

Geoculture was first defined by Wallerstein in 1991 as "a set of ideas, values, and norms that were widely accepted throughout the system and that constrained social action thereafter."[2] Most geographical locations large enough to be politically organized as countries, states, or cities comprise different cultural patterns that may need to coexist and consequently influence one another.

The term geolaw derives from the "law and geography" school of legal thought emerging in the 1980s in the United Kingdom, United States, and Canada. This approach studied the relationship between the space (object) and the norms (actions) in it.[3] In 2007 Goés called geolaw the positivist constitutionalization of geopolitics. In the present globalized world, geolaw is highly influenced by the geopolitics from the centers of global power, as some of them, such as the US, recognize the importance of geoprocessing (a framework and set of tools for processing geographic and related data used to perform spatial analysis)[4] to try to solve social conflicts in an interdisciplinary way.[5]

Furthermore, geoculture and geolaw may be considered soft power mechanisms, able to function even without any conscious intent. However, it is responsible to assume some possible level of intention, especially when coming from the current, well-established, and sophisticated state powers. This is especially true in the assumption of a postmodernist philosophical point-of-view when concepts such as fifth-generation warfare and unrestricted warfare are academically discussed.[6]

The end of the Cold War marked the birth of a new world order, now called postmodern, which came into force with the collapse of the Soviet empire and endures until today. As a philosophical movement it began in the late twentieth century as a clash between two main writings, namely Francis Fukuyama's *End of History and the Last Man* (1992)[7] and Samuel Huntington's *Clash of Civilizations* (1993).[8] Postmodernism is broadly skeptical, suspicious about an objective reality and the quality of human reason and logic.[9] Some of the postmodern world's drastic transformations result from four great moments of paradigmatic rupture in the history of mankind: the fall of the Berlin Wall (1989), the 9/11 attacks on United States soil (2001), the crisis in the liberal financial system (2008), and the coronavirus pandemic (2019).

Hence, many notions could represent postmodernity, such as (1) the distrust of the discourse of modernity metanarratives and their pretensions of timelessness and universality (as may be the understandings about democratic government and financial liberalism), (2) the cultural logic of Fredric Jameson's late capitalism,[10] and (3) the hypermodernity of Gilles Lipovetsky,[11] based on the view that there was no break in the paradigm of modernity and its values: individualism, the explosion of consumerism, political liberalism, and so forth.[12]

Geoculture and Geolaw: Epistemological Convergences Relevant to All States

As previously defined, geoculture and geolaw are fresh concepts derived from the multiple subjects that comprise geopolitical and cultural studies (such as geography, climatology, political sciences, social sciences, etc.) using interdisciplinary approaches and merged as transdisciplinary terms. In essence, they are consequences of epistemological convergences from the subjects considered. And those subjects were considered due to their relevance to state powers in the postmodern world.

In essence, geopolitical postmodernity must be analyzed as a new reality that brings with it a new archetype of hegemonic power relations, very different from its predecessors, namely the previous Cold War and before it the Eurocentric world view prevalent for centuries since before WWII. Geopolitical postmodernity is no longer focused on a Western civilization point of view, as noted by the current increase of geopolitics influences from Indo-Pacific states. Thus, what cannot be denied is that the idea of geopolitical postmodernity brings within it a new concept of vital space (lebensraum), which is the conquest of markets and minds. It is no longer the classic concept of lebensraum linked to the conquest of territories that matters. What matters now is to gain geopolitical muscle to conquer new markets, which are opening on a planetary scale of competitiveness.

Technological advances in transportation reduce the time needed to cross large distances and in telecommunication bring people close. Both advances contribute to reducing the practical meaning of political borders as obstacles against foreign cultural influences. Such boundaries can be rigid, floating, and transient, according to Carla Ladeira Pimentel Águas.[13] Borders can be identified as natural, political, ideological, historical, cultural, legal, geopolitical, and, perhaps, other categories may be named.

The separating boundary has an imaginary or physical dividing line that marks the separation between different spaces. On October 1, 2017, for example, the Catalans took to the streets to claim their independence, breaking the Constitutional Consensus of 1978.[14] Without debating the dichotomous merits of who is or is not right, language, identity, and historical past show the differences between Catalonia, one of Spain's historic nations, and the Spanish state.

The line, therefore, that separates these peoples is imaginary; it exists in the feeling of belonging of each one of these peoples. Strictly speaking, it can be said that there are several historical nationalities in the territory of Spain, such as Catalonia, Basque Country, Galicia, and Andalusia, for example. The autonomous communities are defined by the 1978 constitution to translate, in a way, the diversity of historical nationalities in Spanish territory.[15] It can be seen as an example of geolaw merging several geocultures in a single sovereign state. Many other borders delineated by geoculture show separations between populations of the same city or the same national state.

Mexico belongs to the physical geography of North America;[16] however, in cultural, economic, and geopolitical terms it has traces of Latinity. This Mexican Latinity comes from three main sources: (1) the geographic and cultural location named Latin America, which includes Mexico and the other two American regions of Central America, and South America; (2) its Spanish language, considered as of Latin-derived origin; and finally (3) the ethnicity, which varies along different historical and geographical periods, originally naming European Italic peoples and presently associated to either Latin European peoples from several countries who colonized and populated Mexico (mostly from Spain) or the Latin American peoples, also from several countries besides Mexico.[17]

Brazil congregates several cultures, ethnicities, and peoples under a single federative republican state but without any recognized nations within it. As Brazil occupies about 60 percent of the Amazon rain forest, the largest and most biodiverse in the world, it attracts attention from the international community. And so do the Brazilian native indigenous populations that live not only in the Amazon, but also in demarcated territories across the whole country. About 13.75 percent of Brazilian land territory is demarcated to about 680 indigenous lands.[18] The country acts to integrate those indigenous populations, bringing them food, health, and other supplies while trying to protect their cultures and traditional lands. Those protected lands are rich in several types of natural resources and life forms, and the native cultures remain with so much information to be studied and understood.[19] Studying the native traditional cultures using transdisciplinary approaches may reveal environmental knowledge; new languages and lingos; alternative usages for some natural resources such as new drugs, food, materials; and other insights.

The Brazilian armed forces are important federal instruments to enact this cultural and logistic integration in a beneficial way to all the societal groups involved while keeping the state territory's sovereignty and integrity.[20] FAB is responsible for the aerial integration, a key component, due to the long distances involved and to the difficult access to many indigenous lands that lay in remote regions such as (but not only) the huge Amazon forest. No wonder FAB's motto is "to control, defend and integrate."[21] Furthermore, some of those demarcated territories coincide with Brazilian borders. Nevertheless, integrating those traditional indigenous societies is a matter of national defense to Brazil.

Geoculture Studies

With this as background, a survey of geoculture studies emerges, since the idea of a state is not limited to the simple sum of nations or peoples living on a given territory and under the aegis of the same legal order, but, rather, a living organism that evolves and is linked to indefinite generations that succeed among themselves and whose values continue to share in the present and in the future. As generations pass by, what survives is the self-identification of those peoples as a supreme national group with specific cultural images and behavior patterns that identify themselves as a single community, a supreme country. Geographic boundaries are unable to separate this kind of feeling, and that is a good reason why immigrants, especially in the first generation born in a new country, usually remain psychologically and culturally attached to their original countries.

State or national frontiers can also be defied by military opposition. The notion of a frontier as a front related to war front is unstable: it progresses, stagnates, and retreats according to the strategic moment of the struggle. At the end of combat, borders can be diluted or demarcated. The frontier is inter-place, a term used by the most important figure of contemporary postcolonial studies, Homi K. Bhabha, for encounter, union, creation, cultural exchange, and negotiation.[22] Areas near the borders, places of intersection, are territories for different practices in times of peace: space for diplomacy, business, shared festivities. All these interdisciplinary conversations converge on defense and more specifically, aerospace power, the primary objective of the reflections here presented.

Geolaw, as its very name supposes, is based on legally ordering earthly territories; beyond the border of Earth's atmosphere, state sovereignty rules change. Beyond it lies the largest environment human knowledge has ever comprised: outer space. And this knowledge is oriented to seek a higher international collaboration, as supranational lawfare agreements prevent countries from conquering territories in that domain and intend to build an order of peaceful development to benefit all humankind.[23]

In outer space, a nation can fly spaceships over any country in the world without any legal restrictions. With the United Nations Security Council monitoring and restricting major territorial grabs after World War II, outer space emerges as a prime area of conquest, though this "conquest" can be understood more as gaining the capacity to reach space than to "acquire" it.[24] Dolman wrote in 2002, "Who controls low-earth orbit controls near-Earth space. Who controls near-Earth space dominates Terra (Earth). Who dominates Terra determines the destiny of humankind."[25] This is the new postmodern reality, in stark contrast to the "World Island" view proposed by Sir Halford John Mackinder, a British geographer who wrote a paper in 1904 called "The Geographical Pivot of History" suggesting that the control of Eastern Europe was vital to control of the world.[26]

Accessing space, studying it, commercializing products and services related to space, and using it as a strength multiplier factor of conventional weaponry systems are strategic goals to an increasing number of countries. Geopolitical concepts applied to the use of space by state and private players require new international law agreements relevant to any aerospace power. The capacity of public international law agreements to regulate the use of outer space emerges as a strategic topic in geolaw. Indeed, such international regulation is a founding element in the construction of a Kantian multipolar order.[27]

Space enterprises are usually expensive enough to stimulate multinational partnerships, such as the present collaboration among Russia, the United States, and some other countries. Countries need to deeply discuss and find solutions to newly important space issues such as the growth of megaconstellations of satellites on limited orbital positions, dangerous orbital debris, positioning and testing of weapons in space, and many others, which will keep arising as the human presence in space increases.

COVID-19

Another example of geolaw and geoculture implications comes from the COVID-19 pandemic.[28] The causing virus, severe acute respiratory syndrome coronavirus 2 (SARS-CoV-2), was identified in December 2019 in Wuhan, China. On January 30, 2020, the World Health Organization (WHO) declared the outbreak an international health emergency. And on March 11, 2020, the WHO declared a global pandemic.[29]

It is in this scenario that the geolaw and geocultural effect of COVID-2 emerges, with impacts like those of the pandemic that occurred at the end of the First World War in 1918: the flu mistakenly called Spanish. The new pandemic has changed the way people relate to each other, relying even more on telecommunication means to avoid potentially dangerous physical contact. Physical contact was formally discouraged and broadly restricted by most governments worldwide.

The pandemic affected economic, cultural, political, social relations, labor relations (employer and employee), and the ways of carrying out tasks, intensifying remote work. Many productive means had to adapt to depend less on the human presence, and many markets migrated to cyberspaces such as social media, e-commerce, and online financial systems. Work, entertainment, and social relations also moved more into virtual settings, and most of those changes seem poised to remain in present and future societies.[30]

In a way, differing views on how to deal with the pandemic increase stigmatization of states with high contamination and mortality, making them outcasts of the international community. Arguably, it is possible to suppose that political pressures from abroad in such a health emergency stimulate foreign geocultural influence on a state's internal geolaw, but it still is an issue to be studied. The online reports "Policy Responses to Covid-19"—produced by the International Monetary Fund for each country—exemplify how all countries' actions are judged during the pandemic.[31]

Among the effects of the epidemic in Brazil, the importance of the Unified Health System (Sistema Único de Saúde, SUS) stands out, managed by the state and with an important role assisting the unemployed, street people, street vendors, and ambulant vendors. The SUS acted rapidly and robustly, demonstrating its preparedness before the virus officially arrived in Latin America and taking measures to reduce mortality and severe cases.[32] Brazilian aerospace power, including FAB, contributed by supporting safe transportation of patients and supplies around the huge national territory.[33]

Narratives of Groups, Nations, and States

There is an amalgamation of worldviews and narratives from some nations existing among states. Some nations seek independence, while others do not. Different cultural groups with their own interests exist within all postmodern countries. Uncountable and almost unpredictable temporary events may happen and pose immediate and dramatic impacts to any postmodern state. During the COVID-19 global pandemic, states differed on how to deal with its impacts both among and within states.

It remains clear that the scholar in international relations and law must be able to grasp the epistemological connection that exists among the potential reconfiguration of the world geopolitical order, the cultural values of a given society, and a society's legal-constitutional order. The knowledge required to conjugate information from so many different subjects is necessarily multidisciplinary. And turning it into useful information to benefit national defense and aerospace power is necessarily a transdisciplinary effort.

Therefore, it is clear from everything presented so far that the idea of inter-, multi-, and transdisciplinary scientific approaches reveals that no national state is found disconnected from its internal cultural patterns and from the international order. Internal and external geocultures from states influence and are influenced by national and international geolaw. And a country's level of dependence or interdependence relies on the degree of internationalization of its national life and the intensity of its geopolitics, national defense, culture, and international relations. At this point, we have verified how the concepts of geoculture and geolaw are closely related to state power, aerospace power, and national defense.

Conclusion

As presented in this chapter, there is an urgent need to develop inter-, multi-, and transdisciplinary studies that enable scientific dialogue about geopolitics, culture, and law. One of the great challenges of the postmodern state is to strengthen the epistemological encounter among those three to benefit its own objectives of sovereignty, integrity, and stability. The presented concepts of geoculture and geolaw comprise important, and not fully understood, influences on any state power, including its aerospace component.

It must be recognized that the knowledge about such epistemological convergences is still incipient in Brazil. It is, therefore, a topic under construction, which does not present a definitive conclusion but, rather, reflections and considerations to be debated by the scientific community. Science evolves through dated and situated paradigms that succeed each other, complementing each other dialectically—without presenting, therefore, a full stop. As much as the scientific method does, postmodernism claims there are no apodictic truths and that everything is subject to ongoing reexamination. So, it is right to reaffirm the need for a deeper confrontation of ideas based on facts and evidence previously studied. Science must have no side; it must be objectively supported by data collected and analyzed.

Therefore, the events exemplified in this text have enigmatic geocultural and geolaw developments, including a possible universalization of cosmopolitan human rights in the post–COVID-19 world. The question of the deglobalization of the economy arises also, associated with the probable surge of a multipolar, hegemonic, macro-power architecture, given the geopolitical rise of China and other regional powers, such as India, Russia, and the European Union.[34] This in turn brings up the idea of metaconstitutional multipolarity—that is, a multipolar geopolitical order governed by cosmopolitan constitutional rights of Kantian inspiration. Here is the cornerstone of a true democratic rule of law, which endures in recent times, namely: the dignity of the human person envisioned as a geopolitical-cultural-constitutional axis of democracy.

In conclusion, we defend the idea that, for the first time in human history, the post–COVID-19 world order may be geopolitically multipolar, legally metaconstitutional, and culturally multicivilizational. Scholars should continue to explore and examine the geocultural and geolaw causes and consequences, including the ones focused on aerospace sciences and the Brazilian perspective.

Notes

1. Mafra, *Geopolítica*.
2. Wallerstein, *Geopolitics and Geoculture*.
3. Sanches, "Geolaw and the Geographic-Cartographic Construction as Instrument of Public Politics in the Electric Sector."
4. Esri, "What Is Geoprocessing?"
5. Góes, "O Geodireito e os centros mundiais de poder." Apresentação realizada no VII Encontro Nacional de Estudos Estratégicos, 06 a 08 de novembro de 2007, Gabinete de Segurança Institucional da Presidência da República, Brasília/DF.
6. Abbott, *Handbook of the 5GW*; Liang and Xiangui, *Unrestricted Warfare*; Bartles, "Getting Gerasimov Right"; and White, "Reordering the Law for a China World Order."
7. Fukuyama, *O fim da história*.
8. Huntington, *Clash of Civilizations*.
9. Lyotard, *A condição pós-moderna*.
10. Jameson, *Pós-Modernismo*.
11. Lipovetsky, *Os Tempos Hipermodernos*.
12. Góes, "Geopolítica e Constituição à luz do Estado Democrático de Direito."
13. Águas, "A tripla face da fronteira."
14. Minder and Barry, "Catalonia's Independence Vote Descends into Chaos and Clashes."
15. "La Constitución española de 1978."
16. United States Geological Survey, "NORTHCOM Area of Responsibility."
17. Hale, "Cultural Politics of Identity in Latin America."
18. Ministry of Justice and Public Security, *FUNAI: Autonomy and Indigenous Protagonism*, 49.
19. Meirelles, *O livro de ouro da Amazônia*.
20. Albert, "Terras indígenas, política ambiental e geopolítica militar no desenvolvimento da Amazônia: a propósito do caso Yanomami."
21. Brazilian air force, "Indígena."
22. Bhabha, *O Local Da Cultura*.

23. United Nations Office for Outer Space Affairs, Space Law Treaties and Principles.
24. Klein, *Space Warfare: Strategy, Principles and Policy*.
25. Dolman, *Astropolitik: Classical Geopolitics in the Space Age*.
26. Mackinder, "The Geographical Pivot of History (1904)."
27. Kant, *Perpetual Peace*.
28. Hacisuleyman et al., "Vaccine Breakthrough Infections."
29. World Health Organization, "Coronavirus disease (COVID-19) Pandemic."
30. Caggiano, Castelnuovo, and Kima, "The Global Effects of Covid-19-Induced Uncertainty."
31. International Monetary Fund, "Policy Responses to Covid-19."
32. Croda et al., "COVID-19 in Brazil: Advantages of a Socialized Unified Health System and Preparation to Contain Cases."
33. Santos, "A atuação da Força Aérea Brasileira na operação COVID-19: um estudo de caso."
34. National Intelligence Council, "Global Trends 2040."

Chapter 9

Aerospace Geopolitics

Carlos Eduardo Valle Rosa

Introduction

The world beyond the earth's surface has long captured human imagination. Both the atmosphere and outer space captured attention from scholars of the ancient Greek world: Aristotle, in his work titled *Meteorology*, described climate zones and "developed a primitive model of wind flow on all the known world."[1] Ptolemy undertook an astronomical study in *Almagest, i*n which he dealt with Earth's sphericity and observed the planets' movement; his contribution surpassed surface geography, becoming what would be, for a long time, the scientific comprehension of the cosmos, based on a geocentric model of the universe. Cavalcanti and Viadana point to the contributions of Thales of Miletus, Pythagoras of Samos, Aristotle, and Eratosthenes of Cyrene, the last responsible for "calculating the distance between Earth and the Sun, cataloguing 675 stars, measuring Earth's radius and great circle."[2]

Another such trend occurred between the eighteenth and nineteenth centuries with important founding fathers of modern geography, such as Alexander von Humboldt, who in the "work *Cosmos* described his great view on the Universe";[3] Carl Ritter, who did not limit himself to "description and inventory of objects on the terrestrial surface";[4] and philosopher Immanuel Kant, who observed the totality of natural phenomena and the relationship between space (in geography) and time (in history).[5]

More recently, authors have stepped into the discussion on the extent of the purpose of geography, including Richard Hartshorne, Denis Cosgrove, Peter Adey, Stuart Elden, David Pascoe, and Fraser MacDonald.[6] Even Milton Santos—a prominent Brazilian author covering human geography—makes some allusions to this issue, especially regarding "pioneer zone" and "totality,"[7] when noticing that Earth might reach a "limit-situation" or when writing on the "informational-scientific-technical medium."[8] In this last case, the theorizing may be perceived within aerospace technology (airplanes and satellites), in which the landscape becomes scientific and technical.

In updating this long intellectual tradition, we continue seeking to understand a certain environment under a geopolitical point of view. Whenever the term aerospace is used, attention focuses on a geographic space defined by the conjunction between airspace (or terrestrial atmosphere) and outer space. Beyond perceiving this third dimension from the physiographic point of view that has motivated geography and astronomy since the ancient Greeks, this chapter observes the present and future geopolitical context for that geographic space.

Ever since the balloon's first ascent on a battlefield during the French revolutionary wars around 1793, humans have been aware of the third dimension's geopolitical importance. Stephen Budiansky describes how subordinate officers aboard a balloon would get a broader comprehension of opposing troops' geographic position and layout as well as artillery efficiency, often making them more capable of tactically organizing and directing troop movements than the general in command.[9]

The perspective from above was broadened by pioneer aviators, such as Alberto Santos Dumont, who on October 23, 1906, at the Bagatelle field on the outskirts of Paris, flew for the first time "a minuscule self-propelled airplane, for around a thousand spectators."[10] The airplane not only enlarged the battlefield vision range, going deeper into enemy territory, but was also able to register this vision through photography, attributing new meaning to military intelligence. The First Battle of the Marne, in September 1914, marked a clear Entente victory after French pilot Louis Breguet took photos identifying the tactical movement of the German army heading to Paris.[11]

On October 4, 1957, the Soviet spacecraft *Sputnik-I* opened the possibility of reaching a new tier on the above perspective, the outer space. Cosmonaut Yuri Gagarin was the first man to see Earth from space in 1962, and he considered our planet as "beautiful, with clouds and water, a colorful surface and the darkness beyond," even stating that "Earth is blue!"[12]

From the historical and geopolitical point of view, this continuum formed by airspace and outer space—not only in light of the evolution of technologies and capacities but also from the geographic point of view—configures itself as an aerospace environment and allows for a new form of international power to express itself. This aerospace power contemplates the military side of power plus a wider range of elements, such as the air and space transportation system with its infrastructural networks, civil aviation, aerospace industry, human resources involved in these activities, economy, and aerospace technology, to mention a few aerospace power variables. From a regional cultural perspective, prestige is another valued commodity in aerospace geopolitics.

The trend in English-language literature has been to consider outer space from its own geopolitical perspective, mainly associating this geographic domain to postulates from classic geopolitics. The purpose here is to observe that there is a trend, approached through variables, in characterizing the aerospace environment and to "geopoliticize" this environment, jointly, to constitute an aerospace geopolitics.

Thus, the text is organized around the analysis of five variables that contextualize aerospace geopolitics: the geographic, the political, the economic, the technological, and the ideological. The chapter shall approach only some aspects of these variables, which, certainly, are not the only ones suitable for analysis. However, as they are presented, these variables will establish an overview for this new form of approaching geopolitics from the aerospace dimension.

The Geographic Variable

The geographic element demands the identification of the geographic limit between the terrestrial atmosphere and outer space. This matter is not so evident when factors are considered other than only physical characteristics present in both segments of the aerospace environment. Neither aviation law—especially the Chicago Convention and its regulations—nor space law—the body of regulations from the United Nations Committee on the Peaceful Uses of Outer Space (COPUOS)—clearly defines the transition point between both segments. Currently, this issue is approached by two schools of thought. The first establishes an arbitrary border, known as the Kármán line, and the second, known as the functional approach, considers the laws inherent to physics that regulate the movement or maneuver of devices made by humans for each segment (aerodynamics or astrodynamics).[13]

The geographic variable also involves environmental matters related to the terrestrial atmosphere—and beyond, to planetary surroundings. In response to the impact of carbon dioxide emissions from airplanes, for example, the International Civil Aviation Organization and the International Air Transport Association developed the Carbon Offsetting and Reduction Scheme for International Aviation.[14]

Another example is space debris. In 1982, Daniel Deudney reported on "the 11% yearly rate of growth of space debris."[15] In early 2020, the European Space Agency pointed out the existence of over 120 million objects classified as debris in Earth's orbit, with dimensions of 1 mm to 10 cm, all with the potential to cause damage to spacecraft and generate a cascade effect of catastrophic impact, known as Kessler Syndrome.[16] Obviously, regarding large-sized debris or debris transporting radioactive material, the repercussions would extend to the terrestrial surface. Ilayda Aydin goes further, comparing "space debris to a cancer, with the potential to prevent the use of space by mankind."[17] From the geographic point of view, the great interdependence between the atmosphere and outer space phenomena in relation to third-dimension human activities favors an integrated approach of these geographic spaces.

As a final example of geopolitical manifestation of the geographic variable, consider the competitive aviation market, and expand it. Commercial airline routes and airport slots are used to acquire and expand business and obtain more profits. That is why global partnerships such as the Star Alliance or the Oneworld Alliance explore competitive advantages over their competitors, sharing installations, check-in procedures, customer service, and frequent-flyer programs.[18] For outer space, John Collins had already garnered attention for "strategic locations in space" by discussing the advantages of various orbits and libration points in his book on military geography.[19] Everett Dolman continued this literary trajectory and gave geopolitical meaning to places with geostationary orbits, the Lagrange Points, and the Hohmann transfer orbits in his 2001 book *Astropolitik: Classical Geopolitics in the Space Age (Strategy and History)*.[20]

Expansion by means of new commercial opportunities or natural resources is a common phenomenon in the aerospace dimension. Given this context, it is important to remember the applicability of classic geopolitical concepts such as chokepoints, communication lines, borders, or the vital space to the reality of the aerospace environment.

The Political Variable

The meaning of the word territory in the Greco-Roman world was limited to physiographic aspects.[21] Today, however, territory is a concept that exceeds borders and frontiers, even allowing for culturally focused discussions, as in the case of territoriality, proposed by Robert Sack.[22] The importance of the political variable to the aerospace environment thus is highlighted by the notion of sovereignty.

State sovereignty was first debated in the aerospace field when airplanes surpassed physical obstacles dictated by topography that are commonly associated with a state's territory limits, such as a river, a mountain, or any other geographic element. Lysias Rodrigues, an eminent Brazilian aviator and geopolitician, explained that the territorial border changed due to the aeronautical capacity to pass it.[23] A historical event that justified the perception of this Brazilian geopolitician was the innovative use of the airplane in air control over Palestine, which ensured British colonial administration after World War I. Hugh Trenchard, Royal Air Force Marshal, employed airplanes along the borders of the colonial provinces to observe the movement of insurgent groups and neutralize their maneuvers.[24] Clearly, states' borders and territory control methods underwent a conceptual overhaul in the field of military science.

Alexander de Seversky explicitly assigned a geopolitical meaning to the aerospace environment and aerospace power. His reading of the world space (from a cartographic projection developed on the North Pole perspective) allowed for military geostrategies to develop beyond the classic geopolitical concept of Heartland, proposed by the British land power scholar Halford Mackinder.[25] According to de Seversky, the airplane—and, later, the ballistic missile—would change the comprehension of world geography founded around the traditional perspective derived from the Mercator projection, due to the reach and penetration capacity of air power.[26]

Eyal Weizman, Stuart Elden, Alison Williams, and David Omissi discuss airspace sovereignty, extending the field of power relations among states from the aerospace power perspective, bringing new dilemmas to debates on the reach and boundaries of national sovereignty.[27] Repeated transgressions of territorial sovereignty between Israel and its neighbors, for example, have taken place through airspace.[28] Likewise, the discussion of territory and sovereignty in outer space involves these new aspects. That discussion occurs even as the recently established field of international space law is weakened due to the absence of consensus on some concepts and the lack of clear cases to determine precedents and treaties among the applicable multilateral organizations, such as COPUOS.[29]

As an example, look at the case of Tonga. This country in Oceania has an area of only 747 km² with an economy based on the export of agricultural produce and fishing, as well as tourism. Tonga claimed 16 links of orbit slots at the International Telecommunications Union in 1990.[30] When it took these slots, the country had only one airport with paved runway, a single aviation company, and a single aircraft and no expertise in the aerospace sector.[31] Why would Tonga request slots without the clear capacity to use them?

Another emblematic case is the 1976 Bogotá Declaration regarding geostationary orbits, which determined that "the segments of geostationary synchronous orbit are part of the territory over which the Equatorial states exercise their national sovereignty."[32] When considering that geostationary orbits have a limited amount of room to accommodate satellites, the relevance of this declaration is key, as it indicates an extension of territorial sovereignty from airspace to outer space.

The political variable raises debates in the military field as well, something that has been observed as states weaponize outer space and as antisatellite weapons testing continues, both involving actions starting from the airspace or stations on the surface. Dolman believes that "the militarization and weaponization of space is not only a historical fact, but also an ongoing process."[33] It is important to differentiate space weaponization from militarization. The first phenomenon, still incipient, is about the deployment of weapons in outer space. In turn, militarization emerged from the start of the space race and has hastily intensified ever since. Given that people have weaponized the different geographic domains from the dawn of humanity, it is possible to infer that this process in outer space will be a mere extension of what already takes place in airspace.[34]

A final example of the political variable comes from a recent initiative from the United States government. The executive order of April 2020 seemed to impose interpretation challenges onto international space law regarding the free exploration of celestial bodies.[35] The economic issues regarding asteroid exploration may cause conflicts among nations.

The Economic Variable

The economic variable is important because it provokes disputes over natural resources that are directly associated with economic growth and important to aerospace economy–related state politics.

Even with the economic crisis caused by the pandemic in 2020—which substantially affected not only aviation companies but also whole sectors such as tourism and the aerospace industry—there was a reinforcement of the importance of air transportation concerning basic medical supplies and food for regions severely hit by the pandemic. It must be assumed, as the premise of this study, that the postpandemic period will see a resumption of growth in the aerospace economy, like what was observed in prepandemic years.

According to the Air Transportation Action Group, the aviation market was responsible for more than 65 million jobs in 2017. The Oxford Economics Group states that the aviation market is responsible for 3.6 percent of the world gross revenue.[36] The International Air Transport Association reports that there was a growth of 3.4 percent in air cargo transportation between 2017 and 2018, reaching a total of 64 million tons.[37] Most of this cargo is made up of high-added-value goods and perishables (such as fruit and meat). That reflects the importance of this segment relative to the value of the cargo transported and not in gross terms (tons transported).

The Airports Council International indicates that around 8.8 billion passengers passed through 2,500 airports and 180 countries in 2018.[38] The World Bank provides data on the cargo volume and number of passengers in the air transportation sector, presenting a constant growth curve of these activities each year.[39] In Brazil, the air modal transports from 2010 to 2018 almost doubled in passenger numbers in comparison to the second modal, road transportation.[40]

Space activity follows the same course, as it consists of a "strategic asset," as proposed by Joan Johnson-Freese.[41] Bohumil Doboš points to the importance of resources in outer space as a stimulus for exploring celestial bodies and suggests that the concept of new space introduces new perspectives for the field of space exploration, bringing new players to this economic sector; space activity is increasingly influenced by nonstate actors that have the strength and legitimacy to influence public policies. There are plenty of promising areas in the burgeoning aerospace economy, such as the operation and exploration of launch centers, satellite monitoring services, telecommunication services, natural resource exploration, space tourism, aerospace industries, and the development of new technologies, to name only a few.[42]

A Brazilian example may typify the precise importance of the economic variable. In 2017, Brazil launched the Geostationary Defense and Communications Satellite from the Guiana Space Center in

Kourou, French Guiana (at that time, the Alcântara Launch Center in northern Brazil was not available due to operational problems). According to an Arianespace executive, the cost of launch hovered around $137 million (US dollars).[43] That amount of money could have been saved if the launch center in Alcântara had been working in normal operating conditions.

Regarding the aerospace industry, there are still two points to be considered. The first is that there is a growing tendency to unification; no companies focus exclusively on aeronautical products or space products. All industries of this kind will, probably, be termed aerospace industries. The second point is that the aerospace industry gives rise to scientific innovations and technological progress that benefit society and the economy (not to mention the military). Thus, these industries become more and more important to geopolitical matters.

The Technological Variable

Technology has been vital in the development of geopolitics. Friedrich Ratzel equated a state's technological and scientific development.[44] Alfred Mahan was concerned with maritime technology and the battleship.[45] Halford Mackinder saw the potential for railroads to be influential, through the development of Heartland.[46] All the original theorists of airpower, such as Giulio Douhet and William Mitchell,[47] wrote about military aspects of airpower and how that power could be interpreted in civil aviation.

Today, this approach elevated to outer space can be seen as a sort of new space race, especially when it comes to satellite technology. In fact, as was expressed by Jimmy Teng, "the changes in military technology have shaped the geopolitical landscape since remote times."[48] This quote can also be applied when technology in a broad sense is considered, particularly in aerospace, with a great level of duality in the development process of systems of different natures: almost all big companies in the sector work simultaneously for military and civilian clients.

The impact of technology in aerospace geopolitics involves many approaches. One of them, less common, originates in sociology and anthropology. This disciplinary knowledge leads to understanding an aerial reality (or an *aereality*), a new way of life, or an air life due to the employment of technologies associated with airplanes, airports, and air transportation. Writers such as Saulo Cwerner, Sven Kesselring, John Urry, and Peter Adey, among many others, show the rise of a *homo aeroportis globalis*, a sort of evolution of the *homo sapiens*—a human who always lives on the move through airports and airplanes. This approach corresponds with the view of the airport's influence on city life, not only from the vantage points presented previously but also from the perspective of city architecture, urbanism, transportation networks, and economies.[49]

This type of approach could soon be directed toward the study of outer space—a *homo universum*. Currently, however, the technological impact on outer space geopolitics has been studied on a more pragmatic level. Nayef Al-Rodhan understands "telecommunication and information satellites as the driving force of social change, as the capability of changing political systems."[50] Daniel Deudney refers to the impacts of satellites in astronomy, oceanography, climatology, and geology.[51] Michael Sheehan observes that "space technology bears great benefits for developing nations."[52] That is especially important when national space programs, like the Indian one, are observed. In telemedicine as in agriculture support, India's technological capacity in the space sector has been highlighted in international literature.[53]

Aerospace technology developments often move from industry to society. As an example, the research and products of the National Aeronautics and Space Administration have generated spin-offs such as water filters, computer mice, cellphone cameras, ear thermometers, scratch-resistant ocular lenses, smoke detectors, shoe insoles, air purifiers, and more.[54]

A final aspect of the technological variable connects to the next variable: The Organization for Economic Cooperation and Development revealed that the soft power derived from telecommunications—omnipresent in an age of global information coverage—has taken advantage of aerospace technology to "spread cultural attitudes and political ideas."[55] In this sense, the opportunity to analyze the meaning of ideology in aerospace geopolitics arises.

The Ideological Variable

Our interpretation of ideology refers to the capacity of the state in influencing its people, external actors, and other states. In this approach to ideology, two important elements are described: representation and prestige.

Since the beginning, states have used aviation to express some political goals and an ideology. Discourse surrounding the pioneering of airplanes, and even of balloons, brings national pride to the center of discussions. Famous aerial feats from the first decades of the twentieth century, such as the American Charles Lindbergh and the Portuguese Gago Coutinho and Sacadura Cabral crossing oceans in aircraft for the first time, express a feeling of conquest with national and world relevance. In South America, in 1920, the race between Brazilian aviator Edu Chaves and his Argentinian colleague Eduardo Hearne became an international dispute between Rio de Janeiro and Buenos Aires.[56]

Other examples that lend the air forces of each nation a symbolic sense of representation are the aerobatics (exhibition teams) or the overflights of military aircraft during parades or even at sporting events. In outer space, ideology goes on as a sequel of the same phenomenon observed in airspace. The race to be the first nation to reach terrestrial orbit, send the first man and first woman to space, and land on the moon widened the field of rivalry between the United States and the Soviet Union during the Cold War.

The other element of analysis in this variable is prestige, which is intrinsically connected to representation. As in geopolitics, Colin Gray understands that the geographic space "can speak to the mind and imagination, as well as to eyes and limbs."[57] From this comprehension, aerospace geopolitics should also deal with prestige. Hans Morgenthau saw national prestige as an "indispensable element in rational foreign politics."[58] If the quest for prestige can guide national policies, and if geographic considerations influence such policies, the manifestation that renders national prestige or pride as an important driver in the field of aerospace geopolitics becomes clear.

The manifestation of pride and prestige of a nation still results from a kind of propaganda, "merchandising politics" (propaganda promoted via marketing methods), surrounding the prominence of the aerospace industry, the relationship between an airline company and its country (seen, for instance, in airline companies whose names refer to the country they represent: Air France, Air China, etc.) or the investment in space programs to become leaders in the segment (such as NASA's Artemis program, a mission to send the first woman and first person of color to the moon).[59]

Conclusion

The aerospace environment comprises a connection between airspace and outer space. Considering aerospace power an element intrinsic to national power makes room to include this new geographic dimension in geopolitical studies.

With the intent to geopoliticize the aerospace environment, this chapter opened with the observation of purely geographic elements that would contextualize this environment. This movement allowed a greater immersion in identifying the geopolitical elements that were methodologically proposed in the form of variables.

Thus, if the environment is geographic by nature, and if geography is an essential element of geopolitics, it is appropriate, then, to highlight political, economic, technological, and ideological aspects in a way that allows the suggestion of an aerospace geopolitics. The study did not include other variables that could be studied in the future to supplement the analysis, such as demography (population or migratory studies arising out of the technical evolution), diplomacy (analysis of international agreements or minutes of bilateral meetings, for example), military (which would analyze the force structure, equipment capacity, etc.), sociology/anthropology (study of the social impact of the object of study and the perception of communities on the matter), or biology/environment (analysis of aerial organisms or the climate perspective).

However, the assessments of the selected variables allowed a consistent characterization of a conceptual framework for aerospace geopolitics studies. In the political variable, the issue of sovereignty is emphasized, as it is important in both airspace and outer space. The aerospace sector was shown to be a key economic player despite the 2020 pandemic crisis testing the sector's capacity to recover, particularly the air transportation element. On the other hand, the pandemic also revealed the relevance of air transport to meet the global demand for medical supplies and food.

The aerospace sector is also a driving force of new technologies that have transformed society. This variable also evokes an interrelation with the others when the process of militarization of outer space or the economic potential of remote sensing products is observed. The last theme, the ideological variable, complements the geopolitical meaning of an aerospace environment when it highlights the representation and prestige that have been associated with nations' aerospace power since the early days of aviation and space exploration.

New studies on this proposed geopolitical theme may shed light on aspects yet to be observed. Just as geopolitical theories formulated for surface environments (the terrestrial dimension and water dimension) make no distinctions among the differing terrains (for example, terrestrial geopolitics encompasses plains, highlands, forests, etc., and does not divide them into unique elements), geopolitical studies that consider the aerospace environment as a unified object will pursue an integrated approach between airspace and outer space.

The comprehension of integrality is judged as relevant to the aerospace environment. The future is not perceived as merely aerial geopolitics (of air transportation) or exclusively *astropolitik*. Aware of the differing physical characteristics of the airspace segment and outer space segment, it is understood that there should be, in fact, a comprehensive and integrated aerospace geopolitics.

This chapter is an adaptation of the author's doctoral thesis, defended in 2020, at the Federal University of Rio Grande do Norte.

Notes

1. Bonnett, *What Is Geography?*
2. Cavalcanti and Viadana, "Fundamentos Históricos da Geografia," 11–34.
3. Hugget and Robinson, "Introduction."
4. *International Encyclopedia of Human Geography*, 413.
5. Holt-Jensen, *Geography: History and Concepts*.
6. Hartshorne, *Perspective on the Nature of Geography*; Cosgrove, "Contested Global Visions"; Adey, "Aeromobilities", Adey, *Aerial Life*; Pascoe, *Airspaces*; and MacDonald, "Anti-*Astropolitik*."
7. Santos, *A Natureza do Espaço*, 102, 113.
8. Santos, *Metamorfoses do Espaço Habitado*, 44; and Santos, *A Natureza do Espaço*, 238.
9. Budiansky, *Air Power*.
10. Instituto Histórico-Cultural da Aeronáutica, *História Geral da Aeronáutica Brasileira*.
11. Budiansky, *Air Power*, 56.
12. Becker, "A Amazônia e a política ambiental brasileira," 22–40.
13. ICAO, *Convention on International Civil Aviation*; COPUOS, Report of the Committee on the Peaceful Uses of Outer Space; and Oduntan, *Sovereignty and Jurisdiction in Airspace and Outer Space*.
14. Ahlgren, "The Future of Fuel Source Sustainability with Airlines."
15. Deudney, *Space: The High Frontier in Perspective*, 49.
16. European Space Agency, "Space Debris by the Numbers."

17. Aydin, *Geopolitics of Outer Space*, 33.
18. Hayward, "Airline Alliances."
19. Collins, *Military Geography: For Professionals and the Public*, 146.
20. Dolman, *Astropolitik: Classical Geopolitics in the Space Age*.
21. Smith, *Dictionary of Greek and Roman Geography*.
22. Sack, *Human Territoriality*, 7.
23. Rodrigues, *Geopolítica Do Brasil*, 69.
24. Omissi, "Technology and Repression: Air Control in Palestine 1922–36."
25. Heartland is a "central Eurasian region that traditionally constituted the pivot of world history because of its central location, size, morphology and resource abundance." O'Loughlin, *Dictionary of Geopolitics*, 115.
26. De Seversky, *Air Power: Key to Survival*. The Mercator projection is a cylindrical map projection which became the standard map projection for navigation because it is unique in representing north as up and south as down everywhere while preserving local directions and shapes, especially the countries away from the equator.
27. Weizman, "The Politics of Verticality"; Elden, "Secure the Volume"; Williams, "Hakumat al Tayarrat "; and Omissi, "Technology and Repression."
28. Williams, "A Crisis in Aerial Sovereignty?"
29. Dolman, *Astropolitik*.
30. Buck, *The Global Commons: An Introduction*.
31. Central Intelligence Agency, "The World Factbook."
32. "Declaration of the First Meeting of Equatorial Countries."
33. Dolman, *Astropolitik*.
34. Al-Rodhan, *Meta-Geopolitics of Outer Space*.
35. Pandey and Baggs, "Why Does President Trump Want to Mine on the Moon?"
36. Aviation Transportation Action Group (ATAG), "Aviation: Benefits Beyond Borders."
37. International Air Transportation Association, *Annual Review* 2019.
38. Airports Council International, "Annual World Airport Traffic Report, 2019."
39. World Bank Group, *Air Transport Annual Report 2019*.
40. Ministry of Transport, *Anuário Estatístico de Transportes 2010–2018*.
41. Johnson-Freese, *Space as a Strategic Asset*.
42. Doboš, *Geopolitics of the Outer Space*.
43. Selding, "Former Arianespace Chief Says SpaceX Has Advantage on Cost."
44. Ratzel, "As Leis do Crescimento Espacial dos Estados."
45. Mahan, *The Influence of Sea Power Upon History, 1660–1783*.
46. Mackinder, "The Geographical Pivot of History (1904)."
47. Douhet, *O Domínio Do Ar*; and Mitchell, *Winged Defense*.
48. Teng, *Musket, Map and Money*, 33.
49. Cwerner, Kesselring, Urry, *Aeromobilities*; and Adey, *Aerial Life*.
50. Al-Rodhan, *Meta-Geopolitics of Outer Space*, 35.
51. Deudney, *Space*.
52. Sheehan, *The International Politics of Space*, 126.
53. Nardon, "Developed Space Programmes."
54. Rosen, "No One Should Think that Money Spent on NASA Is a Waste."
55. Organisation for Economic Co-operation and Development, *Geopolitical Developments and the Future of the Space Sector*, 1.
56. Bohrer, *Eduardo Pacheco e Chaves*.
57. Gray, "Inescapable Geography," 161.
58. Morgenthau, *Politics among Nations*, 57.
59. NASA, "Artemis."

Part 3

Aerospace Logistics and Economics

Introduction to Part 3

There is a constant concern in Brazil with issues of a logistical nature, often associated with the development of capabilities in the national aerospace industry. In a country with limited financial resources, the search for effectiveness in processes and greater efficiency in systems is fundamental to the consolidation of aerospace power. Part 3 introduces the themes of logistics and economics, understanding that they are relevant components for understanding this concept in the Brazilian case.

The search for theories that support logistical activities helps implement practices that strengthen aerospace power. Logistics is based on principles that guarantee conditions for force support, and, to a large extent, these principles are supported by empirical experience. However, untested forces cannot prove established principles. Thus, the validation of theories collaborates in the expansion of combat power. A proposed articulation between theoretical bases and the principles of logistics provides a conceptual framework to ensure logical foundations for planning and execution of combat sustainment support. This study of the correlation between theory and principles of logistics is essential for aerospace power as it provides insights for combat logistics, seeking advances in military science and the creation of consistent support solutions from the perspective of war. The study covers various fields such as equipment selection, personnel training, campaign planning, logistics structure, and organizational culture, among other elements of aerospace power.

An important component of aerospace power is related to its logistical capacity. A strong and efficient logistics base ensures the sustainability of aerospace operations and makes resources available in adequate time and quantity. To the theoretical perspective of logistics are added concrete case studies that illuminate another essential element of aerospace power: the economic perspective of the aerospace industry and its technological development.

National defense companies need to master technological capabilities and operate in a favorable economic environment to provide defense equipment and systems that enable success in aerospace power missions. Along with the *National Defense Policy* and other national political-strategic documents, offset is an important practice. Offset is the compensatory practice agreed to between the parties as a condition for the importation of goods, services, and technology, with the intention of generating benefits of an industrial, technological, and commercial nature. Conceived using technological, industrial, and commercial offset clauses in the international procurement processes of the armed forces, the offset has become a public policy whose goal is to consistently favor industrial development and strengthen the aerospace logistic base, both in technological and economic terms. Comparative studies assume relevance as they analyze the strategic objectives and actions presented in national defense documents with the offset practices conducted by the Brazilian Air Force, identifying opportunities for establishing offset agreements as a public policy to provide new capabilities for the Brazilian aerospace industrial base. Considering that this base is an element of Brazil's aerospace power, studies of this nature are extremely important.

The possibility of offset may have been the case with the A-Darter Project, the result of a partnership between Brazil and South Africa for the development of fifth-generation air-to-air missiles. From the Brazilian point of view, international cooperation, especially with partners with similar historical and socioeconomic backgrounds, generates challenges regarding the lack of financial resources, given the basic needs of their populations.

Once again, the idea of a developed and autonomous aerospace industry, with a strong international presence, is reinforced as a fundamental component of the defense logistics complex, not only in the acquisition phases, but throughout the life cycle of air platforms, piloted or autonomous, and their components such as electronic systems and missiles. In this sense, the study of technology transfer processes for the Brazilian aerospace sector, based on international cooperation agreements, is part of the national defense industry's policy of responding to current international security challenges. The A-Darter case reinforces the relevance of the logistical and economic component in Brazilian aerospace power. The cooperation offered the opportunity to access new technologies that can be incorporated into other companies' processes and new products. It also demonstrated that the missile met all the technical performance requirements, resulting in a product with operational performance at the same standard as the best in class available on the market.

The development of new technologies has always characterized the Brazilian aerospace industry, whether in aeronautics or in the Brazilian space program. Unequivocally, this is a relevant facet of the concept of aerospace power in Brazil. There is, effectively, no power in the clear sense of the word without logistical capacity and without an aerospace economy focused on new technologies, development, research, and sustainment capacity.

Part 3, Aerospace Logistics and Economics, reinforces understanding of the scope and multidimensionality of the concept of aerospace power in Brazil. The chapters essentially highlight that logistics, aimed at sustaining the application of force, and economics, aimed at generating force, are integrated with aerospace power in the conception of Brazilian aerospace strategic thinking.

Chapter 10

Logistics Principles and the Axioms of Combat
An Analysis

Luiz Tirre Freire
Fábio Ayres Cardoso

Introduction

To fulfill their constitutional mission in response to demands that the state imposes on internal security and external defense, military forces need to keep their forces operational, from troops to weapons systems, all demanding support. The responsibility for this support rests with the logistics that—through the planning and management of human resources, equipment, and supplies—maintains the effectiveness of the operational structures of the forces, both in headquarters and in campaigns.

Like all other military activities, logistics is based on principles, which, to ensure conditions of practical possibility in real deployment, need to be backed by experience. However, for military forces without combat experience, such as the Força Aérea Brasileira (FAB, Brazilian air force), its set of principles tends to be uncritically emulated by foreign armed forces. Given that countries that are not at war are unable to empirically validate their military logistical principles due to lack of combat experience, these countries should at least seek logical validation of these logistical principles.

Faced with this problem of verifying the logical validity of the FAB's logistical principles in the absence of combat experiences, we argue that theorizing efforts seek links between the principles of logistics and the foundations of combat theory, to be conducted in two sequential actions: (1) to use the combat axioms as a conceptual system, to allow the construction of plausible solutions related to a set of concepts accepted as valid; and (2) to apply the semantic concept of factual reference as a method to identify contexts, classes of reference, and factual predicates that help assess links between the aforementioned principles and axioms.

Analyzing the set of military logistical principles based on the axioms of combat means investigating the set as theoretical object, in which empirical interpretability (derivation of assumptions) and its representativeness (correlation with real properties) can be examined scientifically, assessing conceptual connectivity.

Analyzing the set of logistical principles in the light of combat axioms, through the factual semantics, means identifying reference relationships as well as identifying the occurrence of family predicates from both, defining semantic contexts. Constructs under the same semantic context share the same references and present themselves under the same conjuncture or framework, thus sharing the same cognitive relevance.

Combat Axioms

Axioms are nondemonstrable propositions of a scientific theory used to clearly distinguish their basic concepts, hypotheses, and derivations. The axiomatization process, ordering deductively or inductively known statements on a given subject, is not restricted to mathematics and logic,[1] being used equally in the factual (natural or social) sciences.

Combat axioms were first established by The Military Conflict Institute (TMCI) with the aim of exhibiting, in an orderly manner, key concepts underlying the theory of combat.[2] Insofar as a body of factual ideas comes *ex post facto*, it helps to describe the phenomenon of combat, exhibiting its internal components and dynamics, correlating them in a comprehensive, consistent, and unified way. The concepts of power and potential for combat, missions, unpredictability, and so forth belong to the domain, or semantic reference class, of this theory.

TMCI researchers try to keep this set of axioms to a minimum (Occam's Razor), covering a total of only six, declared and explained below.

Axiom 1: Military combat involves lethal interaction between military forces.

Military forces can be defined as sets of elements activated for the purpose of entering, or threatening to enter, an engagement, for offensive or defensive reasons, in relation to any other set of elements. Lethal interaction includes both the direct use of lethal means and the threat of their use. Although nonlethal weapons can be used, they are always protected by lethal means.

Axiom 2: In combat, each side seeks to achieve an objective (also called a mission) to which a perceived value is assigned.

The mission of a military force corresponds to a specific task assigned to it by a higher authority, or assumed by the commander of the force, based on guidance from a higher body. The mission is a pervasive factor that stimulates and controls combat activity, expanding or restricting its action. It thus acts as a bridge between purpose and result. Combat is, therefore, a purposeful activity within the scope of broader objectives of entities external to the combat itself.

Axiom 3: Combat potential is incorporated into military forces.

Combat potential is the latent ability of a military force employed to achieve results in combat. To create military forces with the potential to wage combat, governments use human, material, infrastructural, informational, organizational, and procedural resources available or capable of mobilization. Nowadays, standards that embody the potential for combat are accepted, such as: structuring forces in hierarchically organized units; establishing control by means of chain of command; admitting lethal and nonlethal armaments; and dividing resources among fire support, maneuver, protection, support, and information activities.

Axiom 4: Commanders on each side activate combat potential with the aim of creating combat power in favor of fulfilling the mission.

Combat power is the instrumental ability of a military force to achieve results in combat. In this sense, each side brings with it energy to create, through combat processes, combat power. Consequently, it is only through these processes that the fighting power can be produced and results obtained. It is emphasized, opportunely, that the results of the combat processes are not determined by each side unilaterally but formed by the interactions between the combatants and with the combat environment.

Axiom 5: The domination of the opposing force is the definitive means to accomplish the mission.

Domination is the result of the imposition of a force's will on the opposing force, through all combat interactions, especially those that affect will and spirit. In combat, this means that one force imposes its will on the other force, with commanders on each side being simultaneously the main transmitters and recipients of domination.

Axiom 6: Uncertainty is inherent in combat.

Uncertainty is a state of doubt about the combat situation, including its outcome. It is ubiquitous and widespread in combat. One of the biggest sources of uncertainty is individual and group behavior under the stress of combat. It manifests its effect on the cognitive states of all combatants and hovers like a fog (fog of war) affecting everyone, especially those in the chain of command. Thus, commanders do their best to reduce the uncertainty on their side while seeking to maximize that on the part of the enemy to obtain a relative informational advantage.

Finally, considering that an axiomatized theory corresponds to the totality of the consequences of its axioms,[3] it establishes, based on assumptions of primitive concepts, the set of basic predicates admitted in its domain.[4] It also helps to construct semantically relevant conclusions between different fields of knowledge that share, totally or partially, similar classes of semantic reference.

Principles of Logistics

The principles of logistics comprise precepts in the form of attributes considered in the planning and execution phases of logistical actions, in favor of successful military operations. They must be integrated into the entire operational structure of the military force, being fundamental to preserve their combat power, expand the reach of strategic objectives, as well as strengthen their ability to sustain themselves in combat.

Although independent, the principles of logistics are also interrelated, requiring from commanders judgment in terms of conditioning to variations in time and space. In view of this, they rarely exert an equal influence on all actions, moments, and places, their relevance and opportunity varying according to the situation. For this reason, commanders need to identify which attributes (principles) have priority in each operation and which will become the foundation for the preparation of their support concept.

The principles of logistics partially differ between military forces, as well as between countries, in that they derive from experiences and lessons learned in real combat. These distinctions also come from the areas of competence or from the specialization of each force in the environments in which they operate. Military logistical principles can be found in both military manuals and academic publications.

In military manuals, the principles of continuity, control, coordination, and security of logistics are widely accepted and employed by national armed forces.[5] Abroad, the variety of military logistical principles is great, produced under different designations, from different classes of references. Some examples of this difference include the principles of agility, reliability, and integration;[6] principles of anticipation, survivability, and improvisation;[7] principles of attainability, simplicity, and sustainability;[8] principles of sustainability and flexibility;[9] principles of cooperation and visibility;[10] and principles of forecasting and balance.[11]

The academic literature on the subject (usually produced by military educational and research institutions, and sometimes independent publications by former members of the armed forces) presents perspectives whose connotative content interposes operational experiences with concepts of logistic theory. In this sense, the principles of synchronization;[12] accountability, shared information, coalition, and civil contracting;[13] dispersion, equivalence, relativity, opportunity, and initial momentum;[14] and cooperation, economics, and forecasting stand out.[15]

The logistics principles established by Brazil's Air Force Command—whose analysis, in light of the combat axioms, is the object of this research—are explained below.[16]

1. Principle of the Objective

This principle deals with directing logistical operations toward a previously defined, intelligible, attainable, measurable goal that contributes to the macro-objective of the military force. The objective guarantees unity of effort as well as focus, helping to prioritize future lines of action as well as the use of the available resources.

2. Principle of Continuity

This principle deals with the uninterrupted provision of support to units at all levels of the war. Continuity guarantees confidence in logistical support, allowing commanders freedom of action, operational reach, and sustainability of their forces. It is obtained through the interdependence between units, a reliable distribution system, and integrated information.

3. Principle of Control

This principle deals with the continual monitoring of the entire logistical process to ensure that the quality and quantity of resources will enable a military force to successfully fulfill its mission. The central issue of control does not mean the exercise of control per se, rather the creation of an atmosphere of moderation and balance to guarantee adherence to the principles of optimized resource management.

4. Principle of Coordination

This principle deals with the union of the individual logistical effort segments—combining both similar and heterogenous elements—of each component of the military force in a single and common effort. Coordination involves administrative processes that ensure commonality, promoting the physical union of the elements existing in the force, to reduce the number of distinct elements.

5. Principle of Economy of Means

This principle deals with the identification and elimination of unnecessary duplications and redundancies, without impairing mission fulfillment. Through economy, resources are provided efficiently, in such a way as to enable commanders to employ all means with the greatest possible effect. It is obtained, thus, using the minimum of resources, within acceptable levels of risk.

6. Principle of Flexibility

This principle deals with the ability to improvise and adapt structures and logistical procedures to changing situations, missions, and operational requirements. Flexibility exists, then, according to the logistical response to an unpredictable environment. It is obtained by anticipating the requirements that will be demanded in an environment in constant change, allowing the development of viable options to support these operational needs.

7. Principle of Interdependence

The principle of interdependence seeks to ensure that combat plans and logistical plans are interconnected—that is, that there is a connection or interdependence between these plans. Interdependence is justified by the fact that the power of the military forces and of the means of combat are encapsulated in resources whose preparation, at all levels of war, is up to logistics.

8. Principle of Objectivity

This principle deals with keeping focus on the objectives of logistical support, in meeting the demands of the fight, and in linking the means and the ends for this service. Objectivity is achieved by focusing on the identification and production of results and the desired final effects.

9. Principle of Opportunity

This principle deals with forecasting, in an attempt to account for the set of circumstances and responding in a timely fashion to make logistical support possible during military operations. Thus, it comprises compliance with the internal clock that determines the time scale of logistical processes, in the provision of resources that will meet operational demands.

10. Principle of Priority

This principle deals with ordering the logistical processes and means that will be used to support the military force, based on criteria of precedence for importance or urgency. Prioritization affects the inflow and allocation of resources to carry out the missions of each combat unit.

11. Principle of Security

This principle deals with the necessary measures to avoid surprise, observation, sabotage, espionage, and restlessness, to ensure freedom of action for logistics. It comprises the axis around which the entire logistics system orbits, being a prerequisite for carrying out its activities in combat situations. It does not imply exaggerated precaution or an avoidance of the calculated risk but rather preserves the planned lines of action.

12. Principle of Simplicity

This principle deals with reducing complexity, an element that introduces confusion in an already chaotic environment like combat operations. Simplicity promotes efficiency in planning and execution and the most appropriate line of action for developing logistical activities, allowing a more effective control of operations. It is achieved with the clarity of the tasks, with standardized procedures, and with clearly defined command relations.

13. Principle of the Command Unit

This principle deals with the unity of authority and program for a set of logistical operations under the same design. The command unit requires a well-defined chain of command, a clear division of responsibilities, an adequate communications system, and a well-understood logistical doctrine, accepted and practiced by commanders at all levels.

The principles of logistics presented are guides to the analytical thinking of air force logistics commanders in evaluating courses of action or support plans. They aim at guiding thought, not specifying actions, and must have their meanings, in terms of semantics, linked to use in combat.

Factual Reference Semantics

Semantics is the area of science that studies the meaning of words, presenting developments in linguistics, philosophy, computer science, and psychology. In factual sciences, it makes use of concepts such as reference, context, meaning, and truth to interpret units of discourse present in reality, from the analysis of the symbol-construct-fact triangle.[17] Thus, semantics can help answer problems related to linguistic facts, such as revealing whether there is a relationship between military logistical principles and combat axioms.

For a semantic analysis to be rigorous, factual (nonconceptual) items like things and events in the real world must be characterized by conceptual items.[18] In the search for precision, unity, and scientific clarity, one should also pursue the mathematical formalism as well as the use of logical metatheoretical constructs and structures as syntactic entities, in favor of systematicity.[19] Under this regime of accuracy, semantics can help elucidate factual referents, as well as estimate their true, factual values.

In this sense (and avoiding regression to the most elementary fundamentals of semantics), some definitions were arbitrarily selected,[20] understood as necessary and sufficient to compose the methodological tool for the analysis of the research. This set is focused on establishing and declaring the main factual semantic referents, such as reference class, context, and correspondence.

Definition 1. T is an axiomatic factual theory if, and only if: (1) T is a factual theory; that is, it contains factual predicates; (2) T is axiomatized; and (3) to every basic concept of T a reference class is explicitly designated by any of the axioms of T.

Definition 2. Being A_i, with $1 \leq i \leq n$, and n the number of axioms of a theory T, the base axiom of T is the conjunction of all axioms of T. That is, there is nothing in the theory that is not in its axioms. Symbolically (A means axiom and df means definition):

$$A(T) =_{df} \bigwedge_{i=1}^{n} A_i$$

Definition 3. The totality F of factual objects is the subset of Ω (set of all objects), formed by nonconceptual elements, in such a way that F is disjoined from C (set of conceptual elements). Symbolically:

$$F =_{df} \{x \in \Omega | \neg(x \in C)\}$$

The actions of fighting and provisioning troops, for example, correspond to factual objects.

Definition 4. If c is a construct, then the class of factual reference of c ($[c]_F$) is the set of factual objects formed by c, where $\Re cx$ is the reference relation between c and x.
Symbolically:

$$c \in C \Rightarrow [c]_F = \overleftarrow{\Re}_F c =_{df} \{x \in F | \Re cx\} \subseteq F$$

Thus, there is a reference relationship ($\Re$) between, for example, the conceptual object "interaction" and the factual object "conflict."

Definition 5: The triple ordinate $\mathbb{C} = \langle S, P, D \rangle$ is called context if, and only if, S is a set of statements in which only predicates of the P

predicate family occur, and the reference class of each predicate **P** in P is included in domain D.

Thus, for axiom 4—symbolically $(\exists x | P_\chi \to p(P_\chi) \to M)$, in which $\exists$ represents exists, x represents the commander; P_χ the generation of combat potential created by x; $p(P_\chi)$ the resulting combat power; and M the mission—all the predicates are in the domain of combat actions, conforming their context.

Definition 6. The reference class of an axiomatic theory is equal to the union of the reference classes of all its axioms.

If
$$A(T) =_{df} \bigwedge_{i=1}^{n} A_i$$

then
$$\mathfrak{R}_F(T) = \bigcup_{i=1}^{n} \mathfrak{R}_F(A_i)$$

Definition 7. Being $\mathfrak{R}(c) = I_1 \cup I_2 \cup \ldots \cup I_f \cup I_n$ the reference class of construct c, in which I_i for i between 1 and $f < n$, are sets of factual items, in such way that $I_i \subset C$ (C is the set of conceptual items). So, the factual reference class of c is the union of the classes of factual items:

$$\mathfrak{R}_F(c) = \bigcup_{i=1}^{f} I_i \subseteq \mathfrak{R}(c)$$

Thus, considering c as a human conflict, in the context of military combat, its class of factual reference may include factual items such as troops, commander, and mission, but not duel, competition, or event.

Definition 8. Two constructs are called coreferential in a context $\mathbb{C}$ $=\langle S,P,D \rangle$ if, and only if: (1) they belong to the context $\mathbb{C}$; and (2) if they have the same reference class. Symbolically, if c and c' are in $\mathbb{C}$, then

$$c \sim c' =_{df} \mathfrak{R}(c) = \mathfrak{R}(c')$$

In this sense, eventually, if the reference class of the constructs present in the logistic principles is a subset, total or partial, of the reference class of the constructs present in the combat axioms, then they are corresponding—being, therefore, semantically linked.

Analysis and Results

Considering the context of "combat support," with S (declarations), P (predicates), and D (domain) inherent to the axioms of combat, a five-step methodological procedure (semantic analysis) was established to identify the existence of co-references between the aforementioned constructs and the principles of logistics. The steps are

- define the constructs present in axioms of combat and principles of logistics;
- present the etymology and the conceptual items derived from each construct established in the previous step;
- identify, from the conceptual items, the factual items related to each construct;
- establish an appropriate factual reference class to represent each construct; and
- compare the classes of factual reference of each set's constructs to identify co-references (semantic binding).

Let c_m, $m = \{1, 2, ..., 6\}$, the constructs that form the axioms of combat, and ℓ_n, $n = \{1, 2, ..., 13\}$, the constructs that form the principles of military logistics. These abstract ideas are based on the main idea of statement of each foundation (step 1). For each c_m and ℓ_n are defined ε_m and ε_n as the etymological origins of each construct, in order to define their original meanings and possible interpretations in combat and logistical contexts. Next, sets of conceptual items κ_m and κ_n exemplify lexicons that share the same semantic field of each construct (step 2). Its objective is to frame different meanings instances (co-hyponyms) and further the identification of factual items I_m and I_n (step 3). Finally, the classes of reference $\Re(c_m)$ and $\Re(\ell_n)$ are derived, representing hypernyms of each construct (step 4), to summarize the meaning of conceptual and factual items in a synthetic expression.

The final results obtained with the execution of steps 1 to 4, for each set of constructs (axioms of combat and principles of logistics), are found respectively in tables 10.1 and 10.2.

Table 10.1. Results of the semantic analysis (axioms of combat)

Constructs (c_m)	Etymologies (ε_m)	Conceptual items (κ_m)	Factual items (I_m)	Classes of reference ($\Re(c_m)$)
c_1: lethality	lethalis (Latin): being lethal, mortal	death, violence, mortal, fight, weapons, danger, cruel, blood, combat, destruction	troops and weapons	armed fight
c_2: mission	missionem (Latin): sending abroad (as an agent)	task, goal, duty, objective, plan, operation, employment, guideline, specialization	plans and orders	purpose (military)
c_3: combat potential	potenciel (Latin): capable of being or becoming combattere (Latin): according to c_1)	power, capacity, intensity, energy, strength, ascendancy, dominance, vigor, dynamics, competence, effectiveness	troops and weapons	destructive capacity
c_4: command	commendare (Latin): order or direct with authority	authority, direction, order, jurisdiction, discretion, government, politics, responsibility	commander	authority (military)
c_5: domination	dominacionem (Latin): rule, control by means of superior ability, resources, or position	pressure, influence, predominance, acting, strength, hegemony, power, defeat, ascendancy, capacity	troops and weapons	power and influence
c_6: uncertainty	opposite of certainty, certonus (Latin): that which is certain, a clear fact or truth	obscurity, doubt, instability, imprecision, confusion, ambiguity	---	unpredictability

Source: authors

Table 10.2. Results of the semantic analysis (principles of logistics)

Constructs (ℓ_n)	Etymologies (ε_n)	Conceptual items (κ_n)	Factual items (I_n)	Classes of reference ($\Re(\ell_n)$)
ℓ_1: objective	*objectivus* (Latin): aim or end of action	goal, purpose, intention, course, direction, target, orientation, intention	director and plan	purpose
ℓ_2: continuity	*continuus* (Latin): uninterrupted connection of parts	continuum, connection, duration, conservation, sequence, progress, constancy	flow of things or activities	persistence
ℓ_3: control	*controllen* (Anglo-French): to check the accuracy of, verify	discipline, regulation, government, direction, management, oversight, restraint	director and plan	management
ℓ_4: coordination	*coordinare* (Latim): to set in order, arrange	organization, plan, systematization, arrangement	director and plan	management
ℓ_5: economy	*oikonomia* (Greek): household, management, thrift	thriftiness, containment, control, moderation, administration	control (of expenses)	management
ℓ_6: flexibility	*flexibilis* (Latin): flexible	agility, resilience, dexterity, tolerance, versatility, capacity, promptness	alternative options	adaptation
ℓ_7: interdependence	related to dependence, *dependentia* (Latin): consequence, result	correlation, reciprocity, compatibility, connection, bonding, linkage, subordination	scheme or project	association

Constructs (ℓ_n)	Etymologies (ε_n)	Conceptual items (κ_n)	Factual items (I_n)	Classes of reference ($\Re(\ell_n)$)
ℓ_8: objectivity	related to the objective (according to ℓ_1)	decision, impartiality, clarity, assertiveness, conclusiveness	clear and direct guidance	purpose
ℓ_9: opportunity	opportunitatem (Latin): fit, convenient time	occasion, chance, convenience, utility, benefit, advantage	schedule	moment
ℓ_{10}: priority	prioratem (Latin): state of being earlier	precedence, primacy, order, rank, prevalence, superiority	preference list	importance
ℓ_{11}: security	securitas (Latin): condition of being secure	safeguard, survival, preservation, insurance, reserve	material guarantees	stability
ℓ_{12}: simplicity	simplicitatem (Latin): state of being simple	openness, uncomplicated, intelligible, elementariness	easy procedure	elementariness and ease
ℓ_{13}: command unit	unitatem (Latin): state of being one. commendare (according to ℓ_4)	singularity, identity, direction, homogeneity, centrality, concentration	director	authority

Source: authors

Considering the contents of tables 10.1 and 10.2, and after analyzing the pairing between each c_m and ℓ_n to find semantic correspondences, it was possible to identify three co-references, the first two being positive and the third being negative:

- c_2 (mission) with ℓ_1 (objective), with regard to bringing together military efforts to achieve an end, has the term *purpose* depicting the class of reference that links the first principle of logistics to the second combat axiom.
- c_4 (command) with ℓ_{13} (command unit), with regard to the act of leading, commanding the execution of actions in combat and being responsible for their results, has the term *authority* de-

picting the class of reference that links the thirteenth principle of logistics and the fourth axiom of combat.
- C_6 (uncertainty) with ℓ_6 (flexibility), ℓ_{11} (security) and ℓ_{12} (simplicity), with regard to actions aimed at dealing or mitigating the confusion and ambiguity (fog of war) present in combat environments.

Positive co-references (letters "a" and "b") present an "affirmative nexus" (agreements) between the classes of reference, maintaining meaning parity, with a logical content of "in favor of" or "same sense." This is largely due to the synonymy of its conceptual items and the sharing of understanding and intention. The negative correlation (letter "c"), on the other hand, expresses "negative nexus" (discordant), in which the classes of references in each group contrast in terms of intention, with logical content "in the opposite direction to." This arrangement follows obeisance to the axioms of combat when it comes to meeting the first five and avoiding the sixth axiom as much as possible.

Regarding the other principles of logistics established by the FAB, it was not possible to identify links to other axiomatized constructs, such as "lethality" (C_1), "combat potential" (C_3), and "domination" (C_5). Thus, unlike other sets of principles of military logistics presented above (which consider, for example, survivability, improvisation, sustainability, attainability and dispersion as principles), this link does not exist in the logistical foundations defined by the FAB.

It is important to emphasize that the axioms of combat were treated as a standard for analyzing the principles of logistics, in which their archetypes would validate them as dedicated to combat support. Thus, it was considered, initially, that this set of axioms are those necessary and sufficient to represent the combat phenomenon. In this way, only insertions or exclusions of logistical principles would ensure the correspondence under study, and not changes in the quantity or content of axioms of combat.

Still, additional inferences can be drawn from the relationship between the logistic classes of reference:

- There are reference classes directly related to management—control (ℓ_3), coordination (ℓ_4), and economy (ℓ_5)—which, despite not having correspondence with the combat axioms, manifest an intrinsic sense to the logistic management itself.

- Objectivity (ℓ_8) is a principle of logistics that proved to be redundant, as it has the same meaning (intention) as the objective principle in its class of reference.
- The principles of economy (ℓ_5), flexibility ℓ_6), and opportunity ℓ_9) are compatible with those established in military manuals of armed forces with combat experience.

The main conclusion obtained from the above results is that the set of principles of logistics established by the FAB, despite showing punctual congruence with logistic principles established by other armed forces, does not adhere to all combat axioms. They are especially aligned with the paradigms of logistic management but neglect the logical links to military employment, such as lethality and combat potential.

Some of the consequences of this attitude of disregarding combative fundamentals in logistics can have major impacts on personnel training processes, equipment selection, campaign planning methods, organizational structures, culture, and logistics management itself—that is, in the force design. This is because the principles of logistics, like any other principles, have the primary function of defining the attributes, properties, or qualities of "logistics things," establishing their identity and ontological status. Military logistical principles dissociated from combat can foster, in the armed forces, the production of planning and execution archetypes aligned with business logistics paradigms, in which predicates of violence, material damage, danger, and risk to life, for example, are absent.

Final Considerations

This work presented a proposal to analyze the principles of military logistics using combat axioms as a conceptual system and the theory of factual semantics as a method. The objective was to support the assessment of the logical validity of the principles of logistics employed by the Brazilian armed forces, given the current lack of combat experiences in Brazil. Initially, the combat axioms developed by The Military Conflict Institute and the set of logistical principles employed by the FAB were presented as a brief literature review. Next, the theory of factual reference semantics was synthesized, presenting concepts of context, factual items, conceptual items, and factual reference class. The method used considered the existing co-references between the existing constructs in the combat axioms and in the principles of logistics, which emerged from the etymological analysis of the constructs and the derived factual items. The main conclusions obtained from the study were that the principles employed by the FAB have a more managerial bias, partially meet the combat axioms—mission and command unit—but do not address the important requirements related to lethality and potential of combat. With this conclusion, we highlight the need to pay attention to the military nature of logistical support and how its identity (attributes) can guide the whole-of-force design and planning.

Notes

1. Bunge, "Why Axiomatize?"
2. DuBois, Hughes, and Low, "A Concise Theory of Combat."
3. Buss, *Handbook of Proof Theory*.
4. Mi and Chen, *Naturalized Epistemology and Philosophy of Science*.
5. Air Force Command, DCA 2-1/2003, *Doutrina de Logística da Aeronáutica*; and Ministry of Defense (MOD), MD42-M-02, *Doutrina de Logística Militar*.
6. Air Force Doctrine Publication 4-0, *Combat Support*.
7. Army Doctrine Publication 4-0, *Sustainment*.
8. Naval Doctrine Publication 4, *Naval Logistics*.
9. Marine Corps Warfighting Publication 3-40, *Logistic Operations*.
10. Royal Canadian Air Force Command, *Force Sustainment*.
11. Australian Army, Land Warfare Doctrine 4-0, *Logistics 2018*.
12. Kress, *Operational Logistics*.
13. Tuttle, *Defense Logistics for the 21st Century*.
14. Huston, *The Sinews of War*.
15. Thompson, *The Lifeblood of War*.
16. Air Force Command, *Aeronautics Logistics Doctrine*.
17. Bunge, *Treatise on Basic Philosophy: Semantics I: Sense and Reference*.
18. Bunge, *Scientific Materialism*.
19. Bunge, "Why Axiomatize?"; and Matthews, *Mario Bunge: A Centenary Festschrift*.
20. Bunge, *Treatise on Basic Philosophy*.

Chapter 11

Offset Practice as a Public Policy
The Case of the Brazilian Air Force

Alexander de Mello Lima
Rodrigo Antônio Silveira dos Santos

Introduction

National power has been defined in the new Brazilian national defense policy proposal as the capability to reach and sustain national interests and objectives. This capability consists of five different elements: political, economic, psychosocial, military, and scientific-technological.[1] In this definition of national power, a solid defense industrial base for Brazil emerges as highly important, connecting with all five elements.

Looking at the military, a solid defense industrial base supplies the armed forces with goods that make it possible to engage in force without relying on foreign suppliers. Thus Brazil can express this instrument of national power with sovereignty and autonomy when compared to other nations. Shifting to the economic aspect, a strong defense industry projects national economy, opens new markets, balances trade, and can also be used to mobilize the country's domestic market, as it is a highly qualified sector and has products with high added value. In the scientific-technological side, the defense industry role stands out as an operator of complex technologies with extremely high reliability levels due to the criticality of their use. Defense products development normally involves cutting-edge technology with potential for dual applicability, putting the defense industrial base as a major driver for national technological development.

In this sense, the defense industrial base figures prominently in Brazilian defense strategic documents, which, in turn, are put into practice by means of governmental action that aims to foster this important industrial sector, mainly regarding technological and industrial autonomy related to defense products. Within this context, one of the most significant governmental actions to benefit companies in the defense sector is the use of technological, industrial, and commercial compensation clauses in the armed forces international procurement processes. This practice, also known as offset, establishes a condition that contracted international suppliers should offer benefits to institutions of the Brazilian defense industrial base to offset foreign currency payments for defense products, as stipulated by the Brazilian Defense Ministry Compensation Policy.[2]

In this context, the present chapter discusses the relationship between Brazil's defense strategy documents and the offset practice during international procurement of defense products and systems. Examining the Brazilian air force's (Força Aérea Brasileira, FAB) offset practice will show how the practice of technological, industrial, and commercial compensation can be understood as a form of public policy to foster such a strategic economic sector.

The Brazilian Defense Strategic Documents as Public Policies

A nation's defense should be understood as a public good provided to society. In this regard, governments must deal with defense issues, such as national security and sovereignty, by means of public policies that foster the development of related institutions and economic sectors.[3]

The first attempt to approach national defense in Brazil as a public policy happened with the edition of the *National Defense Policy* during the presidency of Fernando Henrique Cardoso, in 1996. This policy was updated in 2005 and complemented in 2008 with the *National Defense Strategy* during the presidency of Luiz Inácio Lula da Silva.[4] These two documents will be referred to together in this chapter as the Brazilian Defense Strategic Documents.

Brazilian law establishes that the Brazilian Defense Strategic Documents should be updated every four years. Because of that, they were replaced by an updated version in 2012, during the presidency of Dilma Rousseff. After that, new drafts were proposed to the Brazilian Congress in 2016 and 2020. While the Brazilian Congress was analyzing the 2020 Brazilian Defense Strategic Documents drafts, submitted during the presidency of Jair Messias Bolsonaro, Fernando Azevedo, former minister of Defense, emphasized that "it is not a new policy. The essence is completely the same. As it is a state policy, it is independent of the government, it is cross-governmental. It is practically the same policy of 2012 and 2016 with a few updates."[5]

As stated by Almeida,[6] public policies have the central idea that certain results, desired by society, only have a chance to occur when they are supported by the coercive force or the economic support of the state. These policies are called "public" because they are originated and supported by governments. The state then, as the legitimate executor of force and the protector of people, territory, and national interests, must promote governmental actions to effectively exercise this task. Although they are classified as public policies, the defense documents transcend governmental institutions, which normally seek to provide the expected public services, and permeate various societal sectors such as the defense industry, academia, and the science and technology sector, seeking to direct national power in all its expressions.

Considering its strategic characteristic, the National Defense Policy is a state's public policy, instead of being only a government policy. A state's public policy seeks to address national interests and transcends a specific government, as it proposes political orientation to involve all societal groups, both civilian and military. The orientation given by such a defense policy should not only be embracing along the society, but also must be long-lasting, as one of its most important characteristics is the technological development that looks for autonomy over critical defense technologies. This technological autonomy can only be achieved with years of investment in research and fostering actions that extend past different governments' terms of office.

The importance of technology for national sovereignty makes the National Defense Strategy inseparable from a national development strategy, because a strong defense policy favors a strong development policy. National independence is then achieved by one nation's autonomous technological capability and its own domain of critical capabilities and relevant technologies as a condition either for effective development or for effective national defense.[7] Complementary Law number 97 (1999), which created the Brazilian Ministry of Defense, establishes that national autonomy is related to a nation's socioeconomic development. The armed forces should seek national autonomy by means of nationalizing their means and capabilities, considering research and development to raise national industrial capabilities. The same law presents a subsidiary assignment for the armed forces in which they must cooperate with national development and civil defense.[8]

That said, a nation's defense and development are interdependent. Development is essential to support the military with all needed means, people, and infrastructure.

Brazilian Defense Strategic Documents and the Defense Industrial Base

Given the strategic role of technological autonomy in the national defense context, an effective defense policy should be attentive to the continued strengthening of the defense industrial base. In this sense, the 2020 *National Defense Policy* (*NDP*) spotlights the defense industrial base, declaring that defense budgetary resources should guarantee investments for defense products acquisition, to strengthen the defense industrial base and make possible the development and absorption of technologies to shorten the armed forces' technological gap. It also makes it possible to generate qualified jobs and create conditions to bolster exports.[9]

One of the national defense objectives supported in the 2020 *NDP* is to promote technological and productive autonomy within the defense sector. The *NDP* states that the country should keep and stimulate research and development programs on critical defense technology.[10] It is worthy of note that, since the 2012 version, the *NDP* has included guidance to develop technologies with dual applicability.[11]

The 2012 *NDP* version also labeled the concept of the defense industrial base (DIB), created as an integrated net of public and private companies, civil and military, that perform or conduct research, project, development, industrialization, production, repair, conservation, revision, conversion, modernization, or maintenance of defense products (DP) inside the country.[12] To ensure that this orientation is followed by the whole government, the *NDP* is released in conjunction with the *National Defense Strategy* (*NDS*) since 2012, in such a way that the high-level planning defined in the *NDP* is complemented by the strategic actions that should be taken to reach *NDP* goals. Those strategic actions are presented in the *NDS*.[13]

As the main document to set strategies and actions to increase and maintain national defense, the 2020 *NDS* presented detailed steps to reorganize and strengthen the Brazilian defense industry. One important concept is the triple helix, in which government, industry, and academia synergically interact looking for science, technology, and innovation advancements. This interaction is highlighted as a strategy to address DP needs with the support of critical technologies within national dominance. Out of such interaction provided by the triple helix model, a growth of DIB capabilities is expected, mainly on dual technologies' scope as a way to supply the armed forces with local sustainment and provide the capability to locally develop or modernize DP and Defense Systems (DS) by means of integrated actions taken by governments, universities, and industries to gain technological independence.[14]

The 2020 *NDS* assumes its roots as a public policy and subordinates private interests to the greater public interest. Then, the document says that DIB companies' commercial interests should be subordinated to defense strategic interests. However, it is also important that DIB companies are competitive, to allow exports of DP and dual-use services and technologies.[15]

In search of competitiveness, the *NDS* 2020 states that DIB companies should have access to special legal and tributary regimes to protect them from mercantile immediacy and lack of regularity on state demands for DP, without jeopardizing market competition and development of new technologies. According to the *NDS*, those special regimes must foster conditions to promote higher levels of competitiveness to DIB companies, mainly related to external markets. The idea is to phase in production scale and provide regularity on defense product demand with dual applications. At the end of this process, the expectation is an increase of direct and indirect jobs as a social gain for this public policy and the development of products that will have application either for the military or the civil sector.

In addition to that, the 2020 version of *NDS* also says that the state should keep an acquisition flux that guarantees minimum conditions in which DIB companies can keep their technological and industrial capabilities. This initiative aims to secure the production chain, independently of possible exports and dual products commerce. Besides, the new *NDS* also states that the government should act as a facilitating agent to open new markets by means of an external policy that fosters strategic alliances with other nations. This strategic alliance can focus on industrial cooperation on DP. It shows that a defense policy is a whole government initiative, and it should be linked to the country's external policy as well.

The new version of the *NDS* is structured by defense strategies (DSt), which are then divided into several defense strategic actions (DSA). DSt-8 and DSt-9 address Brazil's national defense objective regarding technological and production autonomy. DSt-8 is related to sustaining the DIB's production chain and deals with financing, research, development, and production of DP, using special legal and supplemental regulations when needed. During the years 2012 and 2013, a legal basis was established with Federal Law 12.598 and Decrees 7.970 and 8.122,[16] which established special norms for DP and DS procurement and development. This legal basis also established a special tributary regime for the Brazilian defense industry as an attempt to reduce tax costs and improve competitiveness inside the defense sector.

DSt-9 aims to strengthen defense science and technology by means of a gradual technology absorption on the national production chain, which should reduce dependency on external technology sources. To implement this strategy, the document emphasizes the triple helix model to synergically promote national development.

The Ministry of Defense Offset Policy

In an increasingly globalized trade and industrial context, it is essential to search for strategic cooperation with other countries on a whole-of-government approach. This cooperation is relevant to acquire technological and industrial capabilities that will lead to national autonomy to produce DP. According to the *NDS* 2020, this kind of partnership should predict that part of development, production, and maintenance of DP happen inside Brazil as a fostering action to DIB companies. Within this context, the Ministry of Defense recommends the use of offset agreements to guide industrial cooperation between Brazilian and foreign companies.

Offsets specifically deal with conditions imposed by governmental contractors over foreign suppliers in which the government uses its purchasing power to search for additional benefits linked to the goods and services that were purchased.[17] Offset practices are recognized by the *Legal Guide on International Countertrade Transactions*, edited by the United Nations Commission on International Trade Law.[18] This document confirms offset transactions as a legitimate commercial practice. According to Transparency International, around 130 countries demand offsets when a foreign company wins a bidding process and signs a procurement contract.[19]

Although offset practices can also be referred to as industrial cooperation, industrial benefits programs, or counterpart policy, offset seeks to provide industrial policy and development. Brazilian federal law defines offset as a practice agreed between the parties related to conditions for the procurement of goods, services, or technology, with the intention to generate technological, industrial, or commercial benefits.[20] In a general sense, offset agreements are a legal tool in which a foreign supplier assumes an obligation with the local government to transfer a set of benefits for the purchasing government. This type of contract normally meets the criteria of conditionality, causality, and additionality.

Conditionality refers to the fact that the offset offer is a requirement put in place by the local government to realize the procurement process from a foreign company. In this sense, it is a condition that is formally communicated with the request for proposal released by the local government in the beginning of the bidding. Associated with this criterion, causality states that the offset offer must be directly related to the goods and services that are being offered by the bidder. In other words, the offered benefits must have a clear link with the goods and services that are the object of the procurement process. Then, the benefits should be a consequence of the needed infrastructure and logistics chain that will be in place to fulfill the commercial contract.[21]

Additionality can be understood under two different aspects. The first refers to the fact that the transferred benefits are offered in addition to the main product—the compensations are not part of the goods or services that are being purchased. Secondarily, additionality represents a technological improvement for the beneficiary institution, with potential to create a new business unit or increase the business volume that already exists in the beneficiary organization.[22]

The current offset policy in Brazil is defined by Normative Ordinance 61/GM-MD, under the Ministry of Defense's responsibility.[23] It is linked to the *NDP* because it was written with the objective of allowing a self-sufficient DP production chain, encouraging it to produce such products in country and generate new businesses to increase national industry competition by means of innovation. This policy also aims to insert Brazilian industry into the international market of defense products and services, which have high added value. Because of that, the document states that offset agreements are under the Ministry of Defense responsibility. However, each armed force has autonomy to set its own basis and organizational structure to centralize offset agreements inside the service and report to the Ministry of Defense. Normative Ordinance 61/GM-MD sets a threshold of US $50,000,000, above which all negotiations related to DP procurement and import shall include an offset agreement. The ordinance also advocates that it is mandatory to include in the RFP the instructions about the offset condition, as well as guidelines to help bidders to present a consistent offset offer.

It is important to clarify that offset agreements do not bring any direct expenses to the commercial contract and the offset costs should be absorbed by the vendor. Because of this condition, the offset agreement accountability is not made with currency values, as it will not have any payment inside the offset agreement. The accountability is made with offset credits to account for the benefits and follow up its implementation. Another important aspect of the offset agreement is that it should have an offset program with additional benefits that fulfill 100 percent of the commercial proposal value, reflected on offset credits accountability.

That said, it is clear the effort to establish the offset mechanism is a strategic tool to strengthen the Brazilian DIB. All the legal basis that has been put in place since 2008 demonstrates that the Brazilian government is looking for possibilities to achieve technological autonomy related to defense products and services. Among those possibilities, offset transactions play a central role to internalize critical defense technology and empower Brazilian companies to develop, produce, and maintain defense products and services in the future.

Offset Practice in the Brazilian Air Force

As already presented in the previous sections, the Brazilian *NDP* 2020 version has the aim, among other objectives, to promote technological and production autonomy in the defense sector. Besides, the Brazilian *NDS* presents the idea that government, industry, and academia should be prioritized and fully integrated to address DP demands based on critical technologies under national dominance. This concept is represented by the triple helix model to assure the needed conditions to absorb, develop, and retain critical technologies.

Under the same principles, the FAB has developed its own offset doctrine, and all procurement processes related to defense products and services import shall result in an offset agreement. FAB has followed the strategy to establish an endogenous capacity for product development and production on Brazilian soil, related with projects from other countries, striving for national aerospace sector autonomy. It should be noted that, to work properly, FAB's strategic option must be made within the scope of a solid public policy. (This strategy differs from the one adopted by other countries, which sought to insert their aerospace production industries into a global supply chain.[24]) FAB has an internal doctrine that was established in 2005 and sets general principles to guide contracting organizations on how to formalize offset agreements. FAB's offset policy is structured around one main objective and seven specific goals that detail organizational interests that should be followed during offset negotiations. The main objective sheds light on promoting technological growth in the Brazilian aerospace industrial park, by means of modern production processes and methods as well as the implementation of innovative dual technologies.[25] The main objective stated above unfolds in different specific goals that emphasize the use of governmental purchase power to benefit the Brazilian aerospace sector, seeking its development by means of the creation of new export markets, the growth of local job markets, the obtaining of external resources to build industrial and technological capabilities, the increase of in-country production inside the defense market, and human resources development and training.

One central aspect of FAB's offset policy is the definition of competencies and responsibilities to practice offset transactions. FAB high staff gives strategic guidance for offset practice to the coordinating organization: the Aerospace Science and Technology Department (Departamento de Ciência e Tecnologia Aeroespacial, DCTA). DCTA's main role regarding offset practice is to concentrate technicians that will provide high-level technical advice for the contracting organizations. Additionally, DCTA will promote integration between FAB and the Brazilian aerospace industry.

In addition to FAB offset policy, an internal regulation called ICA 360-1 details all the steps that should be fulfilled to formalize an offset agreement.[26] This document instructs the precepts for negotiating offset agreements and constitutes the main orientation for contracting organizations.

According to José Augusto Crepaldi Affonso, the decision about offset demand depends on the how the acquisition of a defense product or system will be conducted.[27] Normally, engineering, modeling, and simulation studies are carried out regarding nationalization objectives, logistical support plans, and the preliminary plans to promote development and industrial capabilities. In view of the information generated by these preliminary studies, FAB leadership will select the best option to meet operational needs, whether by purchasing a ready-made product or system, by modernizing a product, or by developing a new product or system. Any of the options could require offset negotiation, and FAB high staff is responsible for defining the general guidelines that will be followed by the contracting organization.[28]

The latest version of ICA 360-1 was published in 2020; it is structured following offset practice phases inside FAB, which are: (1) technological needs identification; (2) requirements' design and issuance; and (3) procurement process and contractual management.[29] Technological needs identification is continuously carried out by FAB, identifying areas of interest, technological needs, and opportunities that could become fulfilled with an offset project. The Ministry of Defense offset policy advises that the benefits of compensation should primarily fill the interests of the contracting force, subordinating these to the strategic interests of national defense. Also, the FAB's strategic planning for 2018–2027 determines the identification of needed technologies to achieve autonomy in critical areas.[30] The identified technological gap can direct an offset requirement that will be negotiated during a purchase process and result in an offset agreement.

According to ICA 360-1, technological needs assessments carried out by FAB seek, primarily, to achieve direct benefits for the air force and, if perceived, possible applications in other armed forces and other government agencies, or even industrial technological developments. The document also provides that FAB's interests must be made compatible with the Brazilian aerospace industrial base's capabilities to successfully absorb the benefits.[31]

ICA 360-1 outlines the following premises to establish an offset agreement:

- Maximum national autonomy in maintenance, operations, and future updates of the purchased products.

- FAB's ability to receive compensation projects as a beneficiary institution, considering human, financial, and material resources for its execution.
- Interest, needs, and armed forces, companies, and other national entities capabilities, to act as a beneficiary institution of compensation projects.

Affonso points out that offset agreements must meet FAB's interests and strategic needs, which should consider the Brazilian aerospace industrial base's capabilities to absorb the benefits, to enable the achievement of practical objectives.[32] The author verified that FAB leadership seeks alignment with the Brazilian Defense Strategic Documents. This attitude demonstrates that FAB's offset policy is strongly directed toward innovation and technological development, meeting *NDP* and *NDS* objectives that are linked to the DIB.

Offset Practice in the Brazilian Air Force as a Public Policy

When comparing the objectives of the *NDP* with the results already achieved by the practice of offset in the FAB—more specifically inside DCTA—it is possible to observe the alignment between FAB's offset policy and the defense strategic documents proposed to the National Congress in 2020.

Among the eight strategic objectives defined in the 2020 *NDP* draft, the third one is related to the promotion of defense technological and production autonomy. *NDP* clarifies that this objective seeks to maintain and stimulate research and development of national technologies in defense critical areas. It also seeks to qualify human capital and develop the DIB and dual technology. Job and income generation are also presented as additional benefits.[33]

The focus of *NDP* is research and technological development to provide the DIB with local capacity to supply the military power. The *NDS* sent to parliament in 2020 points out some strategies to achieve this objective, such as promoting the sustainability of the DIB's production chain and strengthening the area of defense science and technology.

According to Gilberto Mohr Corrêa, 67 offset projects were managed by the Comissão Coordenadora do Programa Aeronave de Combate (COPAC) between 2000 and 2016.[34] COPAC is a FAB organization that manages the most significant acquisition projects and associated offset agreements inside the air force. The study found that 55 projects (82.1 percent of the total) provided an increase in technological capabilities of Brazilian organizations. These 55 projects involved technology transfer from foreign organizations to national institutions, with 37 (67.3 percent) having as main beneficiaries aerospace companies and 9 projects (16.4 percent) having scientific, technological, and innovation institutions receiving the benefits.

The research conducted by Corrêa indicates that, in general, offset practice has already been seeking the same objective in favor of the DIB that was presented in the recent defense strategic documents, because more than 80 percent of the studied offset projects sought to increase the technological capacity to local institutions. It should be noted that even in projects where the beneficiary was a FAB organization, the projects sought autonomy in DP maintenance and operation.[35]

Analyzing in more detail the strategy to promote DIB's production chain sustainment, Defense Strategic Action-47 is exactly the stimulus to obtain commercial, industrial, and technological compensations (offset) related to acquisitions from foreign companies. It can also be observed that within DIB's sustainment strategy, offset has the potential to also cooperate with DSA-45 and DSA-46.

DSA-45 aims to promote exports from the DIB. This action is echoed in one specific objective of FAB's offset policy, in which the creation of new export market opportunities is emphasized. Although this objective has been stipulated since the 2005 version, FAB was not able to fully exploit this course of action, because the predominant offset transaction is the transfer of technology to create national production capability, despite the search to insert Brazilian companies in the international market.

DSA-46, in turn, looks for an increase in local content in DIB products. The current contribution by means of offset practices into this course of action is still limited. For this action to take place, it would be necessary to identify the DIB's production chain and train local suppliers, who generally engage in less specialized and less technologically intensive services, such as surface heat treatments, machining, and composite materials transformation, normally carried out by small or medium companies. The use of offsets might not represent a good alternative to search for this goal.

On the other side, DSA-55, which foresees the use of technological orders applied by the government to increase national content on DP, could be strongly impacted using offset agreements with the inclusion of clauses to provide the progressive fulfillment of nationalization indexes. Offset projects may include subcontracting transactions and work packages that benefit national companies. This type of transaction is extremely important when associated with technology transfer transactions. By placing a national company as a supplier for the product that will be delivered to the government, the foreign vendor's commitment to technology transfer to the national subcontractor is guaranteed. Companies with national capital are more associated with greater technological intensity capabilities than subsidiary companies of foreign vendors. National companies are more related to the provision of services to foreign customers that involve adaptation activities and product modifications, while the subsidiary companies carry out more activities with lower added value.[36]

NDS also states a defense strategy to strengthen the area of defense science and technology (DSt-9), aiming to promote technological and productive autonomy in the area of defense. Within this strategy, FAB's offset practice can contribute to strategic actions DSA-49, DSA-50, DSA-51, DSA-52, and DSA-53. DSA-49 encourages the development of defense critical technologies. This action may be put in place with technology transfer transactions, which are widely used in FAB's offsets. In general, the offset can be used to boost science and technology institutes' development projects, especially those linked to the government. Therefore, offset practice would also collaborate with the DSA-50 by means of the triple helix integration model (government, industry, academy).

Technology transfer in offset projects directly linked to imported DP could be carried out to implement strategic actions DSA-51, DSA-52, and DSA-53, which seek to promote the development of nuclear technology, cybernetics, and space systems, respectively. Possibly, the biggest challenge for this implementation would be to provide the needed capabilities for national science and technology institutes to absorb these technologies.

Conclusion

The Brazilian *NDP* presents the five expressions of national power (political, economic, psychosocial, military, and scientific-technological) and highlights the essential role of the DIB to consolidate economic and scientific-technological expressions, enabling the use of the military expression of national power. Therefore, the state needs to have a public policy that strengthens its military industry to guarantee sovereignty in defense key areas.

A robust DIB supplies the armed forces with nationally produced equipment and systems, making the use of force independent of foreign suppliers. At the same time, the industrial base serves as a catalyst for scientific and technological development, as it operates in areas of critical technology and with highly reliable products, often overrunning this productive technological knowledge to the civilian area.

Keeping the same approach as previous editions of the Brazilian *NDP*, the defense strategic projects proposed to the National Congress in 2020 present objectives, strategies, and actions aimed at strengthening the DIB, especially to sustain technological and production autonomy of defense products and systems.

The DIB strengthening strategy proposed by the Brazilian government must use a variety of industrial development tools, among which one of the most significant for increasing technological and productive capability is the offset practice. The offset concept is fundamental to achieve the third objective of the new *NDP* version: the promotion of technological and production autonomy in defense, aligning strongly with the DS-8 and DS-9 proposed in the *NDS*, targeting DIB's local production chain sustainment and the strengthening of defense science and technology, respectively.

In a very objective way, the DSA-47 seeks a stimulus to obtain commercial, industrial, and technological compensations (offset) during international procurement. This action is a key element, as it contributes to other defense strategic actions that shall be implemented, such as DSA-45, -46, -49, -50, -51, -52, -53. and -55.

Therefore, FAB's offset practice during recent years is a strategic action that contributes to the achievement of the *NDP* objective related to the defense industry. This contribution is greater during technological and production capability increase, since more than 80 percent of FAB's offset projects are related to technology transfer transactions.[37] This fact is perfectly aligned with *NDP* and *NDS* objectives of achieving technological and production autonomy of defense products and systems.

Regarding technological and production capability sustainment within the DIB, it was observed that the contribution of the offset is smaller, by means of only few transactions that encourage exports and the creation of local suppliers for the DIB. In future studies, the causes of this scarcity of export-oriented transactions on FAB's offsets can be analyzed, but one of the hypotheses is the lack of maturity of most national industries in the face of the defense global market.

In conclusion, the insertion of offset clauses in defense products and systems international procurement, such as carried out by FAB, can be seen as a public policy, which, by increasing the technological and production autonomy of the defense industry, is associated with the state policy that targets national sovereignty.

Notes

1. Ministry of Defense (MOD), "Política Nacional de Defesa e Estratégia Nacional de Defesa."
2. MOD, "Portaria Normativa n.o 61/GM-MD."
3. Almeida, "Política de defesa no Brasil."
4. Almeida.
5. Defesanet, "Poder Executivo Entrega Atualizações Da PND, END e LBDN Ao Congresso Nacional."
6. Almeida, "Política de defesa no Brasil."
7. MOD and Secretariat for Strategic Affairs of the Presidency of the Republic, Decree no. 6703.
8. MOD, "Lei Complementar No 97, de 9 de Junho de 1999."
9. MOD, "Política Nacional de Defesa e Estratégia Nacional de Defesa."
10. MOD.
11. Brazilian National Congress, Decreto Legislativo No 373, de 25 de Setembro de 2013.
12. Brazilian National Congress.
13. Almeida, "Política de defesa no Brasil."
14. MOD, "Política Nacional de Defesa e Estratégia Nacional de Defesa."
15. MOD.
16. Civil Office, Decree 12598; Civil Office, Decree 7970; and Civil Office, Decree 8122.
17. Corrêa, "Resultados da Política de Offset da Aeronáutica."
18. UNCITRAL, *Legal Guide on International Countertrade Transactions.*
19. Ribeiro and Inácio Júnior, "Política de Offset Em Compras Governamentais."
20. Civil Office, Decree 12598.
21. MOD, "Portaria Normativa n.o 61/GM-MD."
22. MOD.
23. MOD.
24. Corrêa, "Resultados Da Política de Offset Da Aeronáutica."
25. MOD and Air Force Command, Ordinance 1395/GC4.
26. Air Force Chief of Staff and Air Force Command, "Decree No. 393/GC4."
27. Affonso, "A Política de Offset Da Aeronáutica No Âmbito Da Estratégia Nacional de Defesa."
28. Affonso.
29. Air Force Chief of Staff and Air Force Command, "Decree No. 393/GC4."
30. MOD and Air Force Command, Decree 2102/GC3.

31. Air Force Chief of Staff and Air Force Command, "Decree No. 393/GC4."
32. Affonso, "A Política de Offset Da Aeronáutica."
33. MOD, "Política Nacional de Defesa e Estratégia Nacional de Defesa."
34. Corrêa, "Resultados da Política de Offset da Aeronáutica."
35. Corrêa.
36. Corrêa.
37. Corrêa.

Chapter 12

Transfer of Technology in Military Projects
A Case Study of the A-Darter Project

Carlos Roberto Santos
Patrícia de Oliveira Matos

Introduction

In 2006, the Ministry of Defense of South Africa invited the Ministry of Defense of Brazil to build a strategic partnership to develop a state-of-the-art missile incorporating the most modern technology available at the time. The partnership targeted the joint development of a fifth-generation air-to-air missile, of short-range (between 8 to 12 km). The weapon would be developed following operational requirements defined by the South African Air Force (SAAF) and by the Força Aérea Brasileira (FAB, Brazilian air force), to be used by fighter planes from both countries.

The binational project, which received the name A-Darter (Agile Dart) in Brazil and Assegaai (*also* assegai, a metal-tipped spear used in southern Africa) in South Africa, was in line with guidelines established in the Brazilian *National Defense Policy* (*Política Nacional de Defesa, PND*), which, among other aspects, described the need for a broad technological autonomy for the defense sector, including prioritization of international partnerships allowing a transfer of technology (ToT) process.[1]

The A-Darter also represented opportunities for operational, technological, and commercial gains for all stakeholders in addition to providing greater development for national companies in the field of high-tech missiles,[2] as it would enable Brazilian technicians to train in cutting-edge technologies that are difficult to access in areas such as propulsion, control algorithms, infrared seekers, electronic countermeasures, and inertial sensors.[3] The development responsibility was borne by a South African company, with the participation of the Air Force Command (Comando da Aeronáutica, COMAER) specialists and selected Brazilian companies acting together both in South

Africa and in Brazil, to assimilate all the technology to be developed and to produce the main missile systems/parts in Brazil.[4]

After 17 years, the A-Darter project has concluded its certification campaign and last contractual stages, completing its development cycle. Once the results were felt in both the operational sphere and critical and strategic technologies sphere for the Brazilian defense industry, a study about the A-Darter project was carried out, focusing on the effectiveness of its ToT model.[5] The research delved into available published material and data from the project's reports, which are part of its administrative management process in the file of the Coordinating Committee of the Combat Aircraft Program (Comissão Coordenadora do Programa Aeronave de Combate).[6] Semi-structured interviews were also carried out with representatives of participating companies and FAB officers. Bozeman's Contingent Effectiveness Model of Technology Transfer was used to assess the effectiveness of the ToT model for the technological qualification of participating Brazilian companies.[7]

This chapter is divided into five sections. After the introduction, the concept of technology transfer is discussed, then the Bozeman model for effectiveness assessment. The next section examines the A-Darter project structure, focusing on its ToT model and the joint development process. Then we analyze the A-Darter project through the lens of the Bozeman model through primary and secondary data and information collected during interviews. The final considerations in the concluding section discuss the main findings from the technology transfer process of the A-Darter project as well as the lessons learned from this project.

Technology Transfer and the Effectiveness Model

The term "transfer of technology" or "technology transfer" presupposes transmitting all the knowledge that generated the technology in the first place—that is, sharing the full domain of the technology. This cannot be confused with the simple teaching of an employed technique. Even if the instructions needed to produce goods and services originate in the set of knowledge coming from technology, simply owning these instructions (plans, drawings, specifications, standards, manuals) and having the ability to use them do not mean that user automatically obtains the same complete knowledge as those who originated it.[8] It often appears that many negotiations meant to transfer technology from big multinational companies to developing countries are in fact only training to reproduce techniques for manufacturing products and using equipment.

According to Eduardo Viotti, the essence of ToT is in the national innovation systems found in industrialized countries' economies, which have—besides the ability to absorb existing production technologies—the capacity to incorporate new technological capabilities, allowing them to improve, innovate, and create derived technologies.[9] For technology transfer be considered successful, its recipient (assignee) must be able to effectively use and eventually adapt it, proving its complete understanding, which can be verified by learning by doing, by learning by using, or even by access to technical documents and scientific and technological literature.[10]

K. Ramanathan generically defines ToT as the technology moving process from one entity to another.[11] Rhonda Phillips adds to this by stating that the concept becomes more comprehensive when classifying the technology movement in different domains: from lab to industry, from developed to developing countries, and from one field of knowledge or application to another.[12] From governments' points of view, a technology transfer has the objective of improving living conditions in less developed places or countries through the production of higher-value-added consumer goods. This drives improvement of the human resources used in production and an increase in local business volume.[13]

Edwin Mansfield classifies the ToT process as vertical and horizontal. The vertical ToT is characterized by the technology movement following its several phases: basic research, applied research, development, and production. Conversely, the horizontal technology transfer refers to movement of applied technology from a given context, place, or company to another.[14] The horizontal ToT might have its origin in any phase from the vertical ToT cycle.[15]

This study focused on the horizontal or external ToT, because it is this movement pattern that most closely resembles the technology transfer process that occurred in the A-Darter missile development. In this model, the flow of developed technology from the South African company to participating Brazilian organizations (government, industry, and scientific and technological institutions) was established to enable them to absorb all phases of the missile's development and independently produce and adapt this technology in the future.

To allow technology transfer to take place in the truest sense of the term, the receiver must absorb the full corresponding knowledge in conditions to adapt it according to receiver's particularities, as well as enhance the understanding or even creating a derived technology from the acquired knowledge. However, it is relevant to consider that ToT is a dynamic process, since the technology itself is a dynamic entity: a high-level technology today becomes conventional tomorrow and obsolete on the next day.[16]

Contingent Effectiveness Model of Technology Transfer

Bozeman has idealized a ToT model that starts from a premise that ToT process actors have distinct efficacy criteria and objectives. The term "contingent" indicates, by definition, that ToT includes several parts that have several objectives and, consequently, multiple criteria of effectiveness.[17]

To judge effectiveness, the Bozeman model establishes that the process of analysis encompasses five determining dimensions: transfer agent, means of transfer, transfer object, transfer receiver, and demand environment. The characteristics of the five dimensions are summarized in table 12.1.

Table 12.1. Dimensions of the Contingent Effectiveness Model

Dimension	Focus	Object
Transfer agent	The nature of the institution or organization that wants to transfer the technology, its history, and culture	Government agency, university, company, etc.
Means of transfer	The means, formal or informal, through which technology is transferred	Literature, license, personal exchange, collaborative research, workshops, technical consultancy, spin-off, etc.
Transfer object	The content and form of what is transferred, the transferring entity	Scientific knowledge, physical technology, technological artifact, process, know-how, etc.
Transfer receiver	The organization or institution receiving the object of the transfer	Human, scientific, and technical capital; resources, manufacturing experience, marketing, geographic location, diversity, business strategy, etc.
Demand environment	Factors (market or not) related to the need of the transferred object	Existing demand for the object of the transfer, potential for demand (market-push and market-pull), economic character of the transfer, etc.

Adapted from Bozeman, "Technology Transfer and Public Policy: A Review of Research and Theory."

Bozeman points out that although these dimensions are not entirely comprehensive, they are broad enough to include the main variables in the study of government ToT activities, encouraging the investigation of peculiarities of each case, instead of determining a standard applicable to all situations.[18] Considering model dimensions, the ToT effectiveness assessment is carried out based on seven proposed criteria: out-the-door; market impact; economic development; political rewarding; opportunity cost; human, scientific, and technical capital; and public value.[19]

One of the main assumptions is that none of the effectiveness criteria are considered in isolation. The evaluation method of each criterion is based on answers for key questions collected during interviews or questionnaires applied to main actors responsible for the process, according to the synthesis presented in table 12.2.

Table 12.2. Technology transfer effectiveness assessment model synthesis

Efectiveness criteria	Key question	Objective / Research criteria
Out-the-door	Was the technology transferred?	Exclusively evaluates the output of the transferred technology, without considering the impacts of the activity on the receiver. It is concerned with deadlines and goals.
Market impact	Did the transferred technology result in a commercial impact, a product, profit, or change in market share?	Focusing on intracompany ToT, it evaluates whether there was any impact on the market, measured in terms of commercial success in relation to sales, profits, or market share percentages.
Economic development	Did the technology transfer process have any impact on the economic development of the region where the company is located?	Analyzes the effects on regional or national economic development. It is especially suitable for public ToT (start-ups and spin-offs).
Political rewarding	Did the agent promoting the transfer or the receiving company obtain political gains in the transfer process in relation to their image?	Assesses whether there were political impacts (such as increased investment) arising from the country's participation in the ToT process or recognition of its capacity as a good industrial partner.

Efectiveness criteria	Key question	Objective / Research criteria
Opportunity cost	What was the impact of the transfer on the use of new technologies and local technical solutions?	Examines both the alternative uses of the technical and scientific resources acquired at ToT (laboratories, equipment, training), as well as impacts not expected by the agents involved in the undertaking, such as enabling the nation to ensure its national security mission.
Human, scientific, and technical capital	Has the ToT led to any increase in the ability to use or develop new technologies or technical solutions from the specialists involved?	Considers the impacts of ToT in improving scientific and technical skills, know-how, technically relevant social capital, infrastructures (networks, user groups, etc.), which support scientific and technological work.
Public value	Has the transfer of technology affected (increased or decreased) the good and broad collective values, socially shared?	It analyzes whether or to what extent the primary public value is being affected by the secondary economic value of technology transfer.

Adapted from Bozeman, Rimes, and Youtie, "The Evolving State-of-the-Art in Technology Transfer Research."

Bozeman's model demonstrates that it makes no sense to study the effectiveness criteria—whether in theory or practice—in isolation, since the ToT impact can be perceived by many different approaches: the transferor, the assignee, how the chosen model is being processed, and the transfer object.[20]

The A-Darter Project

Brazil and South Africa have faced similar challenges in recent decades to be able to invest in cutting-edge military technology and to take their place in the restricted group of countries producing high-tech military products. Due to the international arms embargo imposed by the United Nations in 1977, during the apartheid regime, South Africa developed its own product industry for its defense, standing out in the production of missiles, airplanes, and military vehicles.[21] In the 1980s, South Africa's military industry was recognized as one of the more advanced in the world for its technical capacity and project and production skills.[22]

With the scenario changing during the postapartheid period and the reduction of tensions at its borders, the South African defense sector went into crisis with successive budget cuts.[23] Even so, currently, South Africa continues to supply weapons and other military equipment to several countries such as United States (which has purchased armored vehicles to protect military personnel from mines in Afghanistan and Iraq), China, Sweden, and Zambia.[24]

The development of air-to-air missiles in Brazil started in 1976 with the Piranha MAA-1 Project with the objective to replace the AIM-9B Sidewinder used by FAB and adapt them for a surface-to-air version managed by the Aeronautics and Space Institute (Instituto de Aeronáutica e Espaço, IAE).[25]

The MAA-1 missile went through several redesign processes to correct its seeker (target search head) until 2002, when the decision was finally made to buy a new sensor produced by South African company Kentron (now Denel Dynamics), which defined the beginning of a partnership that in the future would be beneficial for both countries. At first, the acquisition of the sensors allowed Brazil to sidestep the technological restrictions imposed on developing countries in a veiled way. Secondarily, the partnership for the development of the A-Darter missile would give South Africa the necessary financial support, and for Brazil the access to this new technology would allow the country to overcome the technological gap in relation to air-to-air missile production.[26]

Historical Context of Brazil–South Africa Cooperation

The relationship between Brazil and South Africa has experienced different phases and changes in foreign policy priorities, alternating periods of negotiations with long periods of indifference. In January 1961, a new phase of approach was initiated, marked by an increase of trade in goods and capital flow, focused on military equipment supplied by Brazilian industry. This period of intense trade lasted until the mid-1980s, when once again Brazil's interest turned to North American and European countries, relegating its African continent partners to the background.[27]

This distancing period began to be reversed after the end of the apartheid regime (1948–1994) and the democracy recovery process in Brazil (from 1985).[28] The resumption of diplomatic and commercial relations began in the Fernando Henrique Cardoso government, with his official visit to South Africa in 1996; his intention was to demonstrate Brazil's interest in expanding Brazilian companies' business on the African continent. This mission, which was the first visit by a Brazilian president to that country, was returned by Nelson Mandela in 1998, when the two countries held several rounds of negotiations on trade relations, culminating in trade agreements in the military and civil areas.[29]

Afterward, routine trading returned between the two countries, symbolizing changing Brazilian foreign policy priorities, which emphasize development focused on multilateralism as new form of South-South cooperation. This new direction of Brazilian foreign policy was founded on forming flexible alliances and partnerships between emerging powers and regional leaders, promoting soft balancing mechanisms as a strategy to ensure greater relevance in discussions of global issues, seen in initiatives such as BRICS, a coalition involving Brazil, Russia, India, China, and South Africa, as well as India, Brazil, and South Africa Dialogue Forum (IBSA).[30]

IBSA provides sectoral cooperation in different areas, including defense. Within this scope, the cooperation agreement signed between Brazil and South Africa was inserted in matters related to defense, considered the initial milestone of the A-Darter project. Among the goals described in the agreement, standouts include those related to defense cooperation promotion, with emphasis on the areas of research and development, acquisition, and logistical support among stakeholders, including the exchange of experiences in science and technology.[31]

At the time the agreement was signed in South Africa, the Brazilian Minister of Defense was accompanied by a delegation composed of defense industry representatives; among the participants were the main members of the future A-Darter program: on the Brazilian side, Avibras and Mectron. On the South African side, Kentron attended, as did the South African Armaments Corporation (Armscor).[32]

More significant agreements were signed, in the scientific and technological cooperation field in 2000 and the Declaration of Intent with the Demand and Procurement Division (DAPD), for cooperation in Research, Development and Technology in the Defense Area, signed in 2003.[33] This research cooperation targeted the feasibility of joint development of a short-range fifth-generation air-to-air missile. This declaration of intent represented the response to the previous invitation, from South Africa to Brazil, to work collaboratively on a state-of-the-art missile-development project.[34]

According to the A-Darter Project Technologies Report, the COMAER evaluated the proposal as a great opportunity for the FAB and the Brazilian defense industry. The evaluation considered that the project, besides representing operational advantages in relation to the missiles used by the FAB up to that moment, would also allow national industry training in the production and future commercialization of a latest generation of missiles.[35] The negotiations were successful, embodied in a memorandum of understanding signed in 2005 and an expense contract, in 2006, signed between Armscor and COMAER.

The Structure of the A-Darter Project

The transfer of technology in missiles is considered strategic and is not normally offered or marketed. The progress in this area is very slow and requires a great deal of effort, normally depending on cooperation with countries that have some degree of knowledge of this technology, as was the case with the Brazil–South Africa partnership for the joint development of the A-Darter missile.[36] The A-Darter missile is a self-defense system to be used by fighter aircraft against possible enemy aircraft, fulfilling operational requirements defined by SAAF and FAB.[37]

Brazil's participation would allow unrestricted access to processes and technologies, as well as to intellectual and industrial property rights related to joint development. To ensure the achievement of these goals, an unusual and innovative structure was established. Brazil has signed four different contracts: an international one, signed between Armscor and Denel for the development and transfer of technology associated with the A-Darter missile; and three national contracts with Brazilian companies (Mectron, Avibras, and Opto Eletrônica), for the reception and absorption of the technology to be transferred.

The Coordinating Committee of the Combat Aircraft Program (Comissão Coordenadora do Programa Aeronave de Combate, COPAC) was responsible for supervising, monitoring, and inspecting subcontractors' activities. To accomplish these tasks, two monitoring and control groups were created, one in South Africa and the other in Brazil.

The A-Darter development was roughly estimated about US$150 million, and Brazil was responsible for 50 percent of the project cost. The project received financial support from the Brazilian Ministry of Science, Technology, and Innovations, with resources from the National Fund for Scientific and Technological Development,[38] due to its features and potentiality of providing innovation and competitiveness to the Brazilian aerospace scientific and industrial community.

The Technology Transfer Model Adopted in the A-Darter Project

With the A-Darter project and the opportunity it represented, the ToT was established as an unfolding of cooperation agreements, and a strategy was conceived to ensure the full transfer of technology. The main target of the planned model was ensuring its effectiveness. The goal would be to enable participating Brazilian institutions to be able to absorb all phases of A-Darter development, enabling them to produce and modernize it independently in the future.

To accomplish the ToT planned model, two Brazilian work teams were created, called Teams A and B, comprising civilian and military technicians and engineers (see fig. 12.1). Additionally, a digital library for the storage of all project data was created. Team A, made up of COMAER experts and members of Brazilian industry, was assigned to Denel Dynamics facilities in South Africa. This team was tasked with missile subsystems development monitoring with South African engineers as well as ensuring that all knowledge was archived in the A-Darter digital library. Team A would also pass all data, documentation, and experiences to Team B, referred to as mirror teams, located in Brazil, distributed at each participating company headquarters.

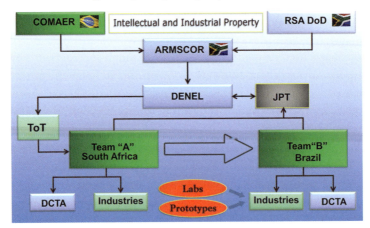

Fig. 12.1. The joint participation of Brazil (COMAER) and South Africa (RSA DoD) regarding Armscor contracting for monitoring Denel's activities in missile development. (Source: Santos, "O Programa Binacional A-Darter.")

Key
COMAER – Comando da Aeronáutica
JPT – joint project team
RSA DoD – Republic of South Africa Department of Defense
ToT – transfer of technology
DCTA – Departamento de Ciência e Tecnologia Aeroespacial (Aerospace Science and Technology Department)

The proposed model predicted that teams installed in national territory should organize the necessary infrastructure (facilities, laboratories, equipment, and software) to be able to reproduce in Brazil the development activities carried out by Denel Dynamics in South Africa. To ensure related subsystems were replicated, items were also manufactured and tested in Brazil following the same procedures as the original project, to prove that the technology transferred was effectively assimilated.[39]

The unusual fact about this ToT model is that it was planned and structured to ensure the transfer of a technology that was not yet available to either country at that time; this means that Brazil and South Africa would develop the technology together, including all associated risks in a complex process like this.

A fifth-generation missile like A-Darter is a complex system that involves many different areas of knowledge, including areas such as optics, fine mechanics, propulsion, embedded computing, composite materials, and digital electronics. Thus, the basic success requirement for a technology transfer process such as the A-Darter is the selection of companies with the necessary technological background and competence to absorb new knowledge—in other words, companies having an appropriate level of maturity in their technical staff as well as an excellent R&D management capacity for new production techniques.

Therefore, subsystems were concentrated in three distinct groups of knowledge and distributed among three Brazilian companies: Mectron was responsible for processing, detection, systems' power, and navigation, as well as being responsible for integrating other systems developed by the other companies; Avibras was responsible for propulsion, detonation, surfaces, and actuators systems; and Opto Eletrônica was responsible for optical systems of missile heads, as shown in figure 12.2.

Critical Factors and Achieved Results

The main technical challenges during A-Darter development came from its infrared detection system, composed of the gimbal assembly (GA) and digital processor assembly subsystems. This system represents the most important and sensitive part of a fifth-generation missile: it is responsible for detecting, identifying, and processing information to intercept the target. Another major technical challenge faced was the rocket engine development and installation, using a modern laser-welding technique associated with the vectored thrust system, adopted to increase the missile's maneuverability.[40]

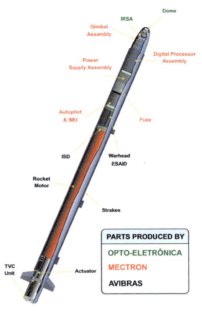

Fig. 12.2. A-Darter missile subsystems and national companies. (Adapted from Denel Dynamics, *A-Darter: Fifth-Generation Air-to-Air Missile System, 2014*. http://admin.denel.co.za/uploads//A-Darter.pdf.)

Key
ESAID - electronic safety, arming, and initiation device
IMU - inertial measurement unit
IRSA - infrared seeker for the A-Darter project
ISD - ignition safety device
TVC - thrust vane control

In addition to these technical challenges, financial crises in Brazil and South Africa led to consequent economic slowdowns, creating difficulties for the fulfillment of financial commitments assumed with the contracted companies. This whole scenario led to a contract extension of more than six years, due to the need to adapt the physical-financial schedule of the project, including all mitigating activities and essential actions to its continuity, such as replanning activities and new equipment configuring for laboratory testing.[41]

With each new difficulty or scenario change, a new round of negotiations was initiated for contract adjustments, enabling project continuity. Throughout the project, six contract amendments were necessary, with varying changes in terms, requirements, and costs, made possible due to the level of confidence achieved between the teams of the two countries.

Along with the project problems, internal company issues also contributed to schedule delays. The management and financing crisis at Denel delayed the industrialization phase for more than a year in South Africa. The decision by Odebrecht, the former owner of Brazilian company Mectron, to dispose of defense business contributed to the unfeasibility of Brazilian companies' participation in the A-Darter production phase, mainly because Mectron would be the integrator of missile systems produced in Brazil.[42]

As a result of these delays, the planned A-Darter integration with FAB F-5EM was canceled and with South African Hawk Mk120 was postponed indefinitely. Nevertheless, the SAAF has completed integration with Gripen, and Denel has been hired to integrate A-Darter with Gripen NG, acquired by Brazil from the FAB FX2 Project.[43]

Even though not everything occurred as planned and the program faced some technical and financial difficulties, causing delays in its conclusion, it is possible to highlight some good results. One of these is the A-Darter's low total cost. It was initially estimated at around US $104 million; the final cost after contracts amendments reached $150 million,[44] divided between the two countries. As comparison, the development of the AIM-9X Sidewinder, which was just an upgrade to the existing AIM-9M version, cost about $850 million.[45]

Some singular factors of the A-Darter project helped fulfill all established development requirements at such a reduced cost. Three factors were key. The first was its lean design team, composed of a core group of about 30 engineers and managers, integrating a total team of about 100 technicians from Denel, Mectron, Avibras, and Opto Eletrônica. Additionally, SAAF and FAB military personnel were dedicated full-time to the project.

The second important factor was the systems engineering approach used, which proved to be effective in reducing risks and maintaining controlled costs. And the third was the extensive use of laboratory test simulation tools, such as Seeker Image and Missile Simulation (SIMIS), developed by Denel Dynamics for testing critical missile systems like GA hardware development, as well as Hardware in the Loop System (HILS) installations, and development of tools for seeker improvement and verification.

The reliability of HILS and SIMIS enabled the project to be developed with fewer missiles needed for equipment development testing and for test campaigns to prove A-Darter's qualification and certification requirements compliance. In total, only 34 missiles were produced for the entire project, with only 18 being launched from the air—a significantly low mark, considering that similar programs normally launch more than 60 missiles for the same task.[46]

The final testing phase of the missile qualification and certification campaign was completed in the end of 2018, at the Overberg Test Range proving ground in the Western Cape of South Africa. The campaign was composed of four guided launches against the Skua high-speed target drone, manufactured by Denel. Each of these guided launches sought to replicate and verify a different type of air combat maneuver scenario. In all scenarios, the SAAF l Gripen C was used as the launching platform.[47]

Denel completed the formal qualification review of the A-Darter missile in August 2019. Its certification followed in September 2019, issued by the Directorate of System Integrity at SAAF and the FAB Institute for Development and Industrial Coordination (Instituto de Fomento à Indústria, IFI).[48] The certification is official recognition that a system meets the technical, operational, logistical, industrial, and safety requirements established by both the FAB and SAAF and symbolizes the end of the project's development cycle.

Although most of the technical data related to the missile is classified, the available information demonstrates that the SRAAM (short-range air-to-air missile) A-Darter is able to compete in the international market, with emphasis on some of its competitive characteristics: structure without gables, providing low drag, with a maximum range exceeding 20 km (12 miles) and maximum speed around Mach 3; rocket engine with low smoke emission, decreasing the probability of visual launch detection by the target; vectored-thrust rocket engine;[49] self-directed with high sensitivity infrared imager, scanning in two wavelengths, including short wavelength infrared and medium wavelength infrared, providing greater target detection capacity; ability to identify and overcome electronic countermeasures (protection against existing flares); lock-on before launch and after launch capability (can be launched already knowing the identified target or can identify it after launch); and target designation by means of radar, helmet, and missile scan function.[50]

According to information made available by the FAB, considering all these features, it is possible that the A-Darter missile provides an operational upgrade when compared to missiles currently operated in

Brazil, since it may provide an increase in offensive capacity and future complete independence in the use of this type of armament.[51]

A-Darter Project Applied to the Contingent Effectiveness Model

All aspects of the five Bozeman's Contingent Effectiveness Model categories of technology were used to investigate whether the effectiveness of the A-Darter ToT process could be identified from the perspective of different technology transfer factors, exploring the vision of each type of agent involved: assignor, assignee, and intervener.

The Transfer Agent

The Denel company, the technology assignor, was hired for technology development and transfer, and its efficiency was rated. The evaluated criteria were: out-the-door; market impact; economic development; political rewarding; and scientific, technological, and human capital. The analysis of these criteria was based on three sources: project reports available at COPAC, articles published by Brazil and South Africa, and interviews with the representative agents in the ToT process.

Out-the-door. The out-the-door criterion was analyzed using Longo's concept of technology: the organized set of all knowledge associated with the development of A-Darter missile systems (know-why) in conditions of applying that knowledge to new product development. The model adopted for ToT horizontal or external flow—contractually established for the transfer of technology developed by Denel Dynamics to participating Brazilian companies—also was considered.[52]

In the view of the technology assignor, the A-Darter technology has been fully transferred under agreed conditions, although the initial term was postponed by more than six years due to problems with some subsystems development. The company's engineers said they were confident that Brazilian technicians had acquired all the skills necessary for reproducing the technology and that they had been able to develop new products based on the knowledge acquired in the development process.[53]

In interviews with the assignees, who were the main stakeholders in the process, they corroborated the assignor's understanding that the technology was transferred within the agreed parameters, as stated by an Avibras representative: "Under the focus of deadlines, objectives and object, the technology was transferred and the contractual agreement object was delivered, taking into account what was expected, which was to obtain a concept demonstrator of the 'rocket engine and ISD (safety ignition device).' "[54]

Market impact and economic development. Regarding the criterion of market impact and economic development, it was only possible to see results for South Africa, since it was the country that started the industrialization process, without the Brazilian companies' involvement. On the South African side, impacts mostly involved supplier companies for missile production. Approximately 140 companies are already involved in missile production, directly or indirectly, as material suppliers or service providers, according to the engineers.[55]

Political rewarding. Political rewarding had noticeable, practical effects for Denel. According to members of its board of directors, the project allowed company projection in South America, opening market opportunities and partnerships in addition to allowing greater confidence from the state to invest resources in projects developed by the company. Analyzing the political rewarding criterion in the condition defined by Bozeman,[56] the partnership with Brazil has demonstrated Denel's ability to collaboratively develop projects with other countries, as well as increased its credibility in relation to transfer of technology to new partners.

Scientific, technological, and human capital. According to South African program managers' reports, relative to the criterion of human, scientific, and technological capital, the results were positive. Although performance forecasting of Brazilian technicians in South Africa was restricted to receiving the transferred technology, there was intense interaction between teams, increasing the technical level for both countries.[57] South African technicians also point out that the good interaction between the project's professionals caused a technical maturation of the teams and changed their way of acting in the company's internal processes. Interviewees mentioned the perceived leveling up in the certification process, due to standardization of procedures assimilated during the collaborative work with the IFI Brazilian technicians for missile certification.[58]

Means of Transmission

In this category, we attempted to investigate the effectiveness of the means, process, or form by which the technology was transferred, based on the risks assumed. To investigate this criterion of effectiveness, representatives of all process players were consulted.

The model adopted for the A-Darter ToT, foreseeing the division of the technical team into two teams, has the characteristics of Viotti's active and passive learning system, once one team was acting directly in the development with the engineers of the assignor company and the other was reproducing the systems in Brazil.[59] Interviews elicited positive and negative points to each system. As a positive point, the gain obtained with the active participation of some Brazilian technicians in the missile development in South Africa was highlighted, enabling participation in all development phases, including overcoming technological development challenges. According to the Brazilian and South African engineers interviewed, this enabled full assimilation of all knowledge involved in the technology of selected missile systems, including limitations of production process.[60]

The interviews demonstrated that this condition allowed greater interaction between teams and increased the level of confidence among technical staff. Mutual trust in the two teams' competence facilitated knowledge transmission and exchange of experiences beyond the initial expectations, causing on some occasions the exchange of roles between actors (assignor and assignee). It was underlined that Brazilian engineers contributed decisively to the solution of some critical problems in the A-Darter development, as in the solution of missile's navigation system algorithms, for instance.[61] This role switching is presented as a standard of effectiveness in the successful ToT Role Shifting Model.[62]

On the other hand, the choice to prove the effectiveness of ToT by reproducing selected systems in Brazil was not unanimous among the air force members of both countries. According to some of them, although passive learning demonstrated the ability to reproduce technology received (know-how), it did not guarantee full control over the technology itself (know-why), nor does it guarantee autonomous development of new products, which could be a weakness of this passive learning model.[63]

Characteristics of the Transfer Object

In this category, criteria of scientific knowledge, physical technology, technological artifact, process, know-how, compatibility, and suitability to the expectations of the technology recipients were investigated, among others. Interviews and published articles on the qualification and certification campaigns show that project participants confirm the complete fulfillment of established requirements for missile development.[64] The four certification launches were successful, and all acceptance requirements established at the beginning of the program were achieved or exceeded.[65]

According to the engineers' interviews, both countries implemented manufacturing change processes. In South Africa, in addition to raising the level of Denel certification, the A-Darter project played an important role in the innovation process within the company. The technology developed for the project is being applied to other products that present significant technological improvements, depending on the concept of incremental innovation. The South African Marlin long-range missile stands out as an example of this technology transfer; its development path was compressed using the knowledge received from the A-Darter.[66] On the Brazilian side, Avibras confirmed the positive influence on the methods, processes, simulations, and testing methodology in the company's products, due to its participation in the A-Darter Project.[67]

In all, the interviews demonstrated the overflow of the technology received for other processes and products within the companies, classified as internal technology transfer.[68] The Opto Eletrônica Company presented a practical example of a spin-off from the defense sector to the space sector: the transfer of technology from the A-Darter missile lens alignment system to the alignment of lenses in satellite cameras.[69]

The Transfer Receiver

This category focuses on the assignees, organizations, companies, and institutions that have benefited from ToT. The investigation used the following criteria: out-the-door; market impact; economic development; political rewarding; opportunity cost; human, scientific, and technological capital; and evolutionary capacity of technology.

Out-the-door. Regarding the transfer of technology from the transferor to the transferee (out-the-door), the interviews demonstrate a convergence of understanding regarding the effectiveness of this criterion. Both Denel and Brazilian companies' representatives were unanimous in stating that transmission of knowledge and its respective absorption were successful.[70] They did cite caveats: the project's financial restrictions limited the number of technicians from Brazilian companies working on the A-Darter development in South Africa, plus there were operational and infrastructure restrictions. For these reasons, efforts were concentrated on the systems to be reproduced in Brazil that were considered priority for the ToT, such as detection and target tracking, data processing, navigation, detonator, rocket engine, actuators, and vectored thrust.

Therefore, for the systems reproduced in Brazil, the technology can be considered fully transmitted, with complete domain absorption and enabling its application in new products.[71] But regarding the technology not included in the prioritized systems, even if all knowledge had been made available, the ToT would have been only partial and limited, since it cannot yet be proven through its application in the development of other products.

Market impact and economic development. In terms of commercial success, within the criteria of market impact and economic development, it is noteworthy that, unlike South Africa, there is no forecast for the A-Darter production starting in Brazil, and there is no commercial agreement between Brazilian companies and Denel for systems supply or joint missile production. In this context, due to the lack of missile production in Brazil, the impact on economic development of the region where participating companies are based cannot be verified. However, if we consider the absorption of technology and the possibility of its application in other processes and products, it can be asserted that there was a positive impact, albeit in this indirect way.[72]

Political rewarding. In the criterion of political rewarding, the companies' statements are coincident when evaluating participation in the project as positive for their image, as it demonstrated their capacity to absorb advanced technologies and participate in multinational projects, enabling them to participate in other new, similar projects. Another indirect effect of this participation was that it facilitated access to financing for their own projects, due to the credibility gained from the financing agencies.

Still regarding the political rewarding criterion, the Avibras representative considers that in fifth-generation missile development where new technologies are incorporated, a refined engineering demand is needed to fulfill the proposed scope by the ToT. This demand has positively improved the company's image, because after being selected to take part in the missile development, the company was forced to seek highly qualified professionals available at the best universities and encourage technological partnerships with other research centers and institutes.[73]

Opportunity cost and human, scientific, and technological capital. Within the opportunity cost and human, scientific, and technical capital criterion, Brazilian companies needed new equipment acquisition and development processes to incorporate into their productive capacity. The positive impacts on human capital were also underscored by interviewees. According to their reports, the professionals involved in the A-Darter development had the opportunity to train themselves with new complex technologies.[74]

Demand Environment

The technology cost and risk factor criteria were analyzed considering this category, focusing on the demand for the transferred object to transferee and consumer market.[75]

The COMAER already made evident the demand for fifth-generation missile technology in the contracting justification text, when considering participation in the project as an excellent opportunity for the FAB and for the national defense industry. The document also points out that the A-Darter project—besides representing operational advantages in relation to missiles currently in use by the FAB—could also position the Brazilian defense industry for production and future trading of state-of-the art missiles.[76]

According to information from the interviewees (corroborated by similar project budgets' available data), the costs involved in the A-Darter project are lower when compared to missile projects with similar technology. Part of the low-cost development success can be credited to the economic similarity between the two countries. Both countries are used to overcoming the challenges of developing complex projects with the minimum financial resources, betting on creative solutions. As the A-Darter demonstrated, the use of simulation systems to prove parameters made it possible to use a reduced number of test missile launches.[77]

Regarding the identified and assumed risks assessment, the feasibility study pointed out high risks for Brazil's engagement in this project. Among them, political and economic instability and the insecurity derived from the fact that the technology was not yet fully developed in South Africa. Certain areas of technology were not mature or advanced enough, increasing risks and impacting development schedules. For example, problems with the detection system (seeker), were responsible for several development delays.

Final Considerations

Brazil and South Africa have many similarities in relation to facing challenges for their development. In addition to troubled recent political histories, both have experienced difficulties investing in defense products, due to the lack of financial resources to meet their populations' basic needs in areas such as health, education, housing, and security. In this context, it is easy to grasp how challenging the development of a missile would be, with its complex technology and restricted access. Yet producing missiles is extremely strategic for any country's projection onto the world stage. The partnership between Brazil and South Africa began when these countries started a process of rapprochement. South Africa did not have all necessary financial resources to continue with its fifth-generation missile project, and Brazil was trying to develop its defense technology and targeting partners capable of providing access to new technologies. Additionally, Brazil also took on risk in committing ongoing financial support to the A-Darter project, given uncertainties about its industries' ability to engage in a highly complex, long-term project.

With all these considerations in mind when reviewing the established ToT model's effectiveness, it is possible to identify the planned structure for the A-Darter joint development: two mirror teams, one acting directly in the country where the technology originated, with possibility of direct participation in the technology development, and the other working in the receiving country, reproducing the systems already developed. This structure is convergent with the Contingent Effectiveness Model of technology transfer, due to emphasis on both individual environments of transferor and transferee and on the larger environment surrounding them. A peculiarity of this structure is to allow a technology flow in both directions, both vertically with the team participating in the development in South Africa and horizontally with the team involved in reproducing the systems in Brazil.

The peculiar ToT model of the A-Darter project has the advantage of enabling the assimilation and mastery of transferred technology, due to the active participation of Brazilian engineers in the development. On the other hand, interviews showed that systems reproduction in Brazil was not totally effective as a way of proving the ToT and full assimilation of related technology to reproduced systems.

It is possible to conclude that in the participating agents' view, the project was successful in its objectives. In addition, the project has provided an opportunity to access new technologies that can be incorporated into other companies' processes and new products. Even though development problems have occurred, this research has shown that all technical performance requirements have been met, resulting in a missile with an operational performance of the same standard as the best ones in its category available on the market.

Few countries are capable of developing complex defense systems like a fifth-generation missile. Thus, it can be inferred that the Brazil–South Africa partnership for the A-Darter development was an excellent opportunity for scientific and technological advancement for participating companies and that the adopted ToT structure proved to be adequate. Good results have been obtained, although some adjustments are needed for future projects of the same magnitude.

Notes

1. Ministry of Defense (MOD), "Política Nacional de Defesa."
2. MOD, *Apêndice V do contrato 001/DCTA-SDDP/2006, Transferência de tecnologia do Projeto A-Darter*.
3. Franchitto and Rebouças, *Relatório sobre Tecnologias do Programa A-Darter*.
4. MOD, *Apêndice V do contrato 001/DCTA-SDDP/2006*.
5. This chapter is derived from a master's thesis approved by the Air Force University Graduate Program in Aerospace Sciences in 2020. Santos, "O projeto binacional A-Darter e o seu modelo de transferência da tecnologia."
6. Although all management administrative processes related to this project have a degree of confidentiality, only aspects related to contractual structures that are in the public domain were analyzed, without adhering to product specifications or other data that compromise the confidentiality inherent to this project. COPAC is an aeronautical command organization that coordinates the development and acquisition of military, combat aircraft, and related systems for the Brazilian air force.
7. Bozeman, "Technology Transfer and Public Policy," 628.
8. Longo, *Tecnologia militar: conceituação, importância e cerceamento*.
9. Viotti, "Ciência e tecnologia para o desenvolvimento sustentável Brasileiro."
10. Cysne, "Transferência de tecnologia entre a universidade e a indústria."
11. Ramanathan, "An Overview of Technology Transfer and Technology Transfer Models," 4.
12. Phillips, *Technology Business Incubators*, 301.

13. Ramanathan, "The Role of Technology Transfer Services."
14. Mansfield, "International Technology Transfer."
15. Ramanathan, "Overview of Technology Transfer," 5.
16. Longo, *Conceitos básicos*, 115; and Kundu, Bhar, and Pandurangan, "Development of Framework for an Integrated Model for Technology Transfer," 35.
17. Bozeman, "Technology Transfer."
18. Bozeman; and Kundu, Bhar, and Pandurangan, "Development of Framework."
19. Bozeman, Rimes, and Youtie, "The Evolving State-of-the-Art in Technology Transfer Research."
20. Bozeman, "Technology Transfer."
21. Silva, "A cooperação internacional em programas de produtos de defesa e seus atores."
22. Deen, "África do Sul continua sendo a maior fábrica de armas do Sul."
23. Silva, "A cooperação internacional."
24. Deen, "África do Sul continua."
25. Paula, "Míssil MAA-1 Piranha."
26. Silva, "A cooperação internacional."
27. Aguilar, "Atlântico Sul: As Relações do Brasil com os Países Africanos."
28. Mattos and Leães, "Relações Brasil-África do Sul."
29. Ministry of Foreign Affairs, "Republica da África do Sul."
30. Soft balancing involves institutional strategies, such as the formation of coalitions, or limited diplomatic agreements, such as BRICS, with the objective of limiting the power of influence of the great established powers. Flemes, "O Brasil na iniciativa BRIC."
31. Goldoni and Ciribelli, "Relações do Brasil com África do Sul e Angola."
32. Silva, "A política industrial de defesa no Brasil."
33. Agreement signed in Cape Town, on March 1, 2000. In effect since July 25, 2003, and promulgated by Decree no. 4824, of September 2, 2003; and Agreement signed in Cape Town, on June 4, 2003, and approved by Decree n. 784, of July 8, 2005. Civil Office, Acordo entre o Governo da República Federativa do Brasil e o Governo da República da África do Sul.
34. Silva, "A cooperação internacional."
35. Franchitto and Rebouças, *Relatório sobre Tecnologias do Programa A-Darter*, 14.
36. Silva, "A cooperação internacional."
37. Franchitto and Rebouças, *Relatório sobre Tecnologias do Programa A-Darter*.
38. MOD, *Apêndice V do contrato 001/DCTA-SDDP/2006*.
39. MOD.
40. MOD.
41. Information taken from the documents that justify the additive terms to the contract with Armscor (001/CTA-SDPP/2006) TA 04/2011. MOD, "Termos Aditivos ao contrato com a ARMSCOR."
42. Silva et al., "Cooperação Brasil-África do Sul na produção do míssil A-Darter."
43. Olivier, "A-Darter Programme Reaches Maturity."
44. Computed value considering the six additive terms to the contract with Armscor (001/CTA-SDPP/2006). MOD, "Termos Aditivos ao contrato com a ARMSCOR."
45. Olivier, "A-Darter Programme Reaches Maturity."
46. Olivier.
47. Jayme, "Míssil A-Darter conclui testes."
48. Maria, "Evento marca encerramento do ciclo de desenvolvimento do projeto A-Darter."

49. Thrust-vectored rocket engines generate acceleration so an aircraft or missile can be directed, giving them great maneuverability in the air.

50. Jayme, "Míssil A-Darter Conclui Testes."

51. Air Force, "Notícias Sobre 'A-DARTER.' "

52. Longo, "Tecnologia militar."

53. Lt Col Danie du Toit (CEO, Denel Group), in discussion with Carlos Roberto Santos, Brasília, September 26, 2019.

54. Marco Aurélio Almeida (executive sales engineer, AVIBRAS), in discussion with Carlos Roberto Santos, Brasília, September 26, 2019.

55. Japie Maré (A-Darter project manager, Denel Dynamics) and du Toit in discussion with Carlos Roberto Santos, Brasília, September 26, 2019.

56. Bozeman, "Technology Transfer and Public Policy."

57. Maré, discussion.

58. Maré.

59. Viotti, "Ciência e tecnologia para o desenvolvimento sustentável Brasileiro."

60. Almeida, Henrique Pazelli (project manager, Opto Space & Defense), Anderson Mendes Moreira (colonel engineer, Brazilian Air Force), du Toit, and Maré in discussion with Carlos Roberto Santos, Brasília, September 26, 2019.

61. Du Toit, discussion.

62. Kundu, Bhar, and Pandurangan, "Development of Framework."

63. Interviews with a lieutenant colonel of the South African Air Force, du Toit, and a colonel engineer at FAB Anderson Mendes Moreira, in discussion with Carlos Roberto Santos, Brasília, September 26, 2019.

64. Defense Industry Daily, "South Africa, Brazil's A-Darter SRAAM Hits Target."

65. Jayme, "Míssil A-Darter Conclui Testes."

66. Olivier, "A-Darter Programme Reaches Maturity."

67. Marco Aurélio Almeida (executive sales engineer at AVIBRAS), in discussion with Carlos Roberto Santos, Brasília, September 26, 2019.

68. Bach, Cohendet, and Schenk, "Technology Transfer from European Space Programs."

69. Bach, Cohendet, and Schenk.

70. Henrique Pazelli (project manager, Opto Space & Defense) and Marco Aurélio Almeida (executive sales engineer at AVIBRAS), in discussion with Carlos Roberto Santos, Brasília, September 26, 2019.

71. Cysne, "Transferência de tecnologia entre a universidade e a indústria."

72. Pazelli, discussion.

73. Almeida, discussion.

74. Almeida and Pazelli, discussion.

75. Bozeman, "Technology Transfer and Public Policy."

76. Franchitto and Rebouças, "Relatório Sobre Tecnologias Do Programa A-Darter."

77. MOD, *Apêndice V do contrato 001/DCTA-SDDP/2006*; and Almeida, du Toit, and Mare in discussion with Carlos Roberto Santos, Brasília, September 26, 2019.

Conclusion
Carlos Eduardo Valle Rosa

The twenty-first century has reinforced the importance of the third dimension for humanity. The domain of the aircraft reached to outer space—involving the mastery of satellites, rockets, and spacecraft—composing a new geographical space where aerospace power is exercised. This work sought to demonstrate that aerospace power should not be understood only as a power of a military nature, but rather as a comprehensive concept that also includes civil aviation, industry, infrastructure, the technological complex related to this power, and the primordial human resources focused on aerospace activities.

This perception has accompanied Brazil's national strategic thinking since the 1950s, not only in doctrinal documents or in the legislation that shaped this understanding but also in several concrete initiatives. For this reason, it helps to have a tailored volume that encompasses the various dimensions of aerospace power and the Brazilian air force (Força Aérea Brasileira, FAB). Contributions presented in this book cover economics; the administrative and logistics sphere; cultural bias from the perspective of joint operations; military capabilities, including cybernetics; the law applied to military operations; geopolitical understanding; and more.

Part 1, Aerospace Power Applications, included ways in which aerospace power can be observed in everyday practice, focusing on issues such as aircraft employment, the cyber issue, and command and control as represented by civil and military air traffic control. The Embraer A-29 Super Tucano aircraft and unmanned aerial vehicles are discussed in chapter 1, "Small Remotely Piloted Aircraft Weaponization," which allowed a better understanding of risk analysis in military planning when including this equipment in certain scenarios. Cyberspace is addressed in chapter 2, "Cyber Defense in the Brazilian Air Force," in which the structuring and importance of the aeronautical cyber defense center as a sector body of the air force was revealed. In chapter 3, "SIPAM and SIVAM Projects," the authors discussed the challenge of incorporating into systems the idea of command and control in multidomain scenarios and in a modern hybrid warfare environment. Chapter 4, "Brazilian Air Traffic Flow Management," demonstrated that the national commitment to air traffic coordination and control extends to other South American countries to provide more efficient and safe air traffic.

Part 2, Aerospace Power and Contemporary Issues, centered on current issues in which the participation of aerospace power is suitable, such as the issue of international humanitarian law and peace operations and the concept of joint operations, as well as geopolitical approaches on how aerospace power impacted the COVID-19 issue and how aerospace geopolitics is understood. The issue of humanitarian law was addressed in chapter 5, "Brazil and International Humanitarian Law," which concluded that Brazilian defense documents have adhered to the international legal precepts in force, providing legal basis for military actions outside the national territory and strengthening the country's vision of cooperation in multilateral forums. In chapter 6, "The Brazilian Air Force in UN Peace Operations," the authors observed postulates such as respect for freedom, the search for peace, and the contribution to human rights, reflecting on the Brazilian participation in this type of international military operation. In chapter 7, "Interoperability among Brazilian Armed Forces," the issue of joint operations from a cultural perspective was analyzed, seeking to better understand the multidimensionality of the concept of aerospace power. Chapter 8, "Geopolitics, Culture, and Law," discussed the role of aerospace power in the face of geopolitical tensions of the capitalist triad inspired by Locke and the multipolar world reflected in Kant's ideas. In chapter 9, "Aerospace Geopolitics," the geopolitical relevance of the aerospace environment was analyzed through geographic evidence and political, economic, technological, and ideological variables.

Part 3, Aerospace Logistics and Economics, addressed issues that impact the aerospace industry, the scientific and technological complex, and the sustainability of aerospace power. Chapter 10, "Logistics Principles and the Axioms of Combat," discussed logistics, resonating on postulates of combat theory and semantic theory. Chapter 11, "Offset Practice as a Public Policy," analyzed offset practices made feasible in the FAB's purchase processes abroad from the perspective of the aerospace economy. In chapter 12, "Transfer of Technology in Military Projects," integration with strategic partners in the field of technological development of aerospace weapon systems was studied through the analysis of the A-Darter project.

This multidimensionality characteristic of aerospace power in Brazil favors the more effective preparation and use of military power with a stake in achieving national objectives stipulated at the political level. Of course, this way of addressing aerospace issues increases their degree of complexity. The military strategy world, which witnessed the insertion of airplanes in the phenomenon of war, today cares about the growing militarization of outer space. The field of economic and social relations features complexity. The disruption in mobility of people resulting from the COVID-19 pandemic has clearly demonstrated how dependent we are on air transport. The demand for global connection, in real time, reinforces the reliance of today's society on space assets, therefore making it more interdependent and complex to understand our reality.

The main objective of this work was to aggregate interpretations of how to understand the complexity of aerospace power. By passing along important arguments defended by the authors—linked to the Postgraduate Program in Aerospace Sciences (Programa de Pós-Graduação em Ciências Aeroespaciais, PPGCA) of the university of the Brazilian air force—it is possible to foresee some challenges that arise for the study of this interesting theme.

The integration between the scientific and technological complex and the aerospace industry is a subject of great relevance. Embraer's C-390 Millennium project is a good starting point. Designed as a multimission aircraft, it incorporates great mission flexibility (aeromedical evacuation, search and rescue, firefighting, and humanitarian assistance) and low operational cost. F-39 Gripen, a state-of-the-art multimission fighter, reinforces the partnership between the international (Saab) and the institutional (Embraer) for the development of high-tech platforms, including data fusion, electronic warfare, network-centric warfare, intelligent digital cockpits, and a wide range of weapon systems, including missiles beyond visual range.

Highly qualified human resources have been of major importance for the growth of national technological capacities and is a subject of great repercussion in PPGCA and in the FAB. The development of the scramjet 14-X S seeks the Brazilian membership in the select group of nations with hypersonic propulsion capacity, whose speed is more than 6,000 km/h. Similarly, the Link-BR2 Project aims at communicating, in real time and in a safe way, between aerial vectors and command-and-control stations, placing aircraft in shared situational awareness.

In the administrative field, themes that involve the improvement of internal processes, like management and governance practices, are highlighted. These themes are driven by recent structural changes in the Air Force Command, which stimulate military transformation movements (also known as a revolution in military affairs) observed in several countries with significant aerospace capability.

One of these movements reinforced space activities as the responsibility of the national aerospace power. The New Space—a posture that privileges independent entrepreneurial activity based on private actors—testifies to the intensification of commercial space activities. Alcântara Launch Center (Centro de Lançamento de Alcântara, CLA) opening for commercial activities through the technological safeguard agreement with the United States, the public calls for companies interested in making launches from the CLA, and Brazil's recent agreement to NASA's Artemis Accords are just the spearhead for the country's insertion in this reality.

In line with this reality of transformation, greater management capacity of the national space assets is intended. The implementation of the space operations center already has brought significant results, as in the tracking of the space object Atlas 2AS Centaur R/B. This initiative is the first step toward the capability of situational awareness of the spatial domain, relevant due to Brazil's inclusion in an international scenario of extreme strategic importance.

All these issues connected to outer space open the way for legal debates, economic approaches, geopolitical integrations, and development of military capabilities, topics of great relevance to Brazilian aerospace power. It is expected that, after reading this work, the reader has understood the conception of aerospace power used in Brazil, in addition to facing the most exciting challenges against it. For this reason, this conclusion does not end this work, but it should arouse interest in aerospace power studies from a multidimensional, comprehensive perspective, the way it was inserted in Brazil since its origin.

Abbreviations

AAV	assessment and advisory visit
ACC	Area Control Center
AF	armed forces
AFIS	Aerodrome Flight Information Service
AGA	airfields and ground aids (aeródromos e auxílios terrestres)
AI	artificial intelligence
ANAC	Agência Nacional de Aviação Civil (National Civil Aviation Agency)
ANS	air navigation service
APA	análise pós-ação (post-action analysis)
APP	Aeronave Remotamente Pilotada (Guarani Approach Control)
AST	Acordo de Salvaguardas Tecnológicas (Technological Safeguards Agreement)
ATAG	Aviation Transportation Action Group
ATC	air traffic control
ATFM	air traffic flow management
ATM	air traffic management
ATS	air traffic services
BRICS	Brazil, Russia, India, China, and South Africa
CAEPE	Curso de Altos Estudos de Política e Estratégia (Advanced Studies in Politics and Strategy Course)
CAN	Correio Aéreo Nacional (National Air Mail)
CARSAMMA	Caribbean and South American Monitoring Agency
CCOPAB	Centro Conjunto de Operações de Paz do Brasil (Joint Center for Peace Operations)
CCSIVAM	Comissão Coordenadora do Projeto Sistema de Vigilância da Amazônia (Commission for the Coordination of the Amazon Surveillance System Project)

CDCAER	Centro de Defesa Cibernética da Aeronáutica (FAB Cyber Defense Center)
CDM	collaborative decision-making
CENSIPAM	Centro Gestgor e Operacional do Sistema de Proteção da Amazônia (Management and Operational Center of the Amazonian Protection System)
CERNAI	Comissão de Estudos Relativos à Navegação Aérea Internacional (Commission for Studies Relating to International Air Navigation)
CGNA	Centro de Gerenciamento da Navegação Aérea (Air Naviation Management Center)
CINDACTA	Centro Integrado de Defesa Aérea e Controle de tráfego Aéreo (Integrated Center for Air Defense and Air Traffic Control)
CISCEA	Comissão de Implantação do Sistema de Controle do Espaço Aéreo (Airspace Control System)
CLA	Centro de Lançamento de Alcântara (Alcantara Launch Center)
COMAE	Comando de Operações Aeroespaciais (Aerospace Operations Command)
COMAER	Comando da Aeronáutica (Air Force Command)
COMDCIBER	Comando de Defesa Cibernética (Cyber Defense Command)
COMPREP	Comando de Preparo (Preparation Command)
COPAC	Comissão Coordenadora do Programa Aeronave de Combate (Coordinating Committee of the Combat Aircraft Program)
COPE	Centro de Operações Espaciais (Space Operations Center)
COPUOS	United Nations Committee on the Peaceful Uses of Outer Space
COSDEA	Comando de Seguridad y Defensa del Espacio Aéreo (Airspace Security and Defense Command)
DAPD	Demand and Procurement Division

DCTA	Departamento de Ciência e Tecnologia Aeroespacial (Department of Aerospace Science and Technology)
DECEA	Departamento de Controle do Espaço Aéreo (Airspace Control Department)
DIB	defense industrial basis
DICA	Direito Internacional Dos Conflitos Armados (International armed conflict law employment handbook in the armed forces)
DIRCM	directional infrared counter measures
DOS	denial-of-service attack
DP	defense products
DPA	digital processor assembly
DS	defense systems
DSA	defense strategic actions
DSt	defense strategies
DTCEA	Destacamento de Controle do Espaço Aéreo (Air Space Control Detachment)
EANA	Empresa Argentina de Navegación Aérea (Argentine Air Navigation Company)
EGN	Escola de Guerra Naval (naval war college)
EMCFA	Estado Maior Conjunto das Forças Armadas (Joint Chief of Armed Forces State)
END	*Estratégia Nacional de Defesa* (*National Defense Strategy*)
EPI	equipamento de proteção individual (individual protection equipment)
ESG	Escola Superior de Guerra (war college)
ETIR	Equipe de Tratamento e Resposta a Incidentes em Redes Computacionais (Incident Treatment and Response Team in Computer Networks)
EW	electronic warfare
FAB	Força Aérea Brasileira (Brazilian air force)

FARC	Forças Armadas Revolucionárias da Colômbia (Revolutionary Armed Forces of Colombia)
FGS	Force Generation Service
FMP	flow management positions
FMU	Flow Management Unit
FPDAM	Flight Procedure Design and Air Space Management
FUNAI	Fundação Nacional dos Povos Indígenas (National Foundation of Indigenous Peoples)
FURG	Federal University of Rio Grande
GA	gimbal assembly
GAC	Grupo de Acompanhamento e Controle (Accompaniment and Control Group)
GDCS	Geostationary Defense and Communications Satellite
GDP	gross domestic product
GREPECAS	Grupo Regional de Planificación e Implementación del Caribe y América del Sur (CAR/SAM Regional Planning and Implementation Group)
GI	*guerra irregular* (irregular war)
GSI	Gabinete de Segurança Institucional (Institutional Security Cabinet)
GST	general systems theory
HILS	hardware in the loop system
HUD	heads-up display
IAC	international armed conflict
IAE	Instituto de Aeronáutica e Espaço (Institute of Aeronautics and Space)
IBAMA	Instituto Brasileiro do Meio Ambiente e dos Recursos Naturais Renováveis (Brazilian Institute of Environment and Renewable Natural Resources)
IBSA	India, Brazil, and South Africa Dialogue Forum
ICA	Instrução do Comando da Aeronáutica (Instruction of the Aeronautical Command)

ICAO	International Civil Aviation Organization
IFI	Instituto de Fomento à Indústria (Institute for the Promotion of Industry)
IFP	instrument flight procedures
IHL	international humanitarian law
IRT	international relations theory
ISD	ignition safety device
JSOU	Joint Special Operations University
LEA	Línguas Estrangeiras Aplicadas às Negociações Internacionais (Foreign Languages Applied to International Negotiations)
LCC	low-cost carriers
LSC	Laboratório de Sistemas Computacionais (Computer Systems Laboratory)
MAER	Ministério da Aeronáutica (Ministry of Aeronautics)
MINUSTAH	Mission des Nations Unies pour la stabilisation en Haïti (United Nations Mission for Stabilization in Haiti)
MOD	Ministério da Defesa (Ministry of Defense)
MONUSCO	Missão Nações Unidas de Estabilização no Congo (United Nations Mission for Stabilization in the Congo)
MOU	memorandum of understanding
MTW	maximum takeoff weight
MUH	medium utility helicopter
MWIR	medium wavelength infrared
NAFTA	North American Free Trade Agreement
NASA	National Aeronautics and Space Administration
NASIC	National Air and Space Intelligence Center
NAV	Serviços de Navegação Aérea (Air Navigation Services)
NDP	*National Defense Policy*
NDS	*National Defense Strategy*

NGO	nongovernmental organization
NIAC	non-international armed conflict
OECD	Organization for Economic Cooperation and Development
ONU	Organização das Nações Unidas (Organization of the United Nations)
ONUC	Organização das Nações Unidas no Congo (Organization of the United Nations in the Congo)
OOTW	operations other than war
OSD	Operational Support Department
OST	Outer Space Treaty
PAG	*processo administrativo de gestão* (administrative management process)
PAROS	Prevention of an Arms Race in Outer Space
PCRS	Peacekeeping Capability Readiness System
PKO	peacekeeping operations
PND	*Política Nacional de Defesa* (*National Defense Policy*)
PPGCA	Programa de Pós-Graduação em Ciências Aeroespaciais (Postgraduate Program in Aerospace Sciences)
PPWT	Prevention of the Placement of Weapons in Outer Space and the Threat or Use of Force against Outer Space Objects Treaty
PROSUB	Programa de Desenvolvimento de Submarinos (Submarine Development Program)
RPA	remotely piloted aircraft
RSA	Republic of South Africa
RVSM	reduced vertical separation minimum
S&T	science and technology
SAAF	South African Air Force
SAE	Secretaria de Assuntos Estratégicos (Secretariat of Strategic Affairs)

SAM	Saab Aeronáutica Montagens Ltda.
SAMIG	South American Implementation Group
SARP	sistema de aeronave remotamente pilotada (remotely piloted aircraft system)
SCADA	Sistema de Supervisão e Aquisição de Dados (Data Supervision and Acquisition System)
SFA	seeker functional area
SGDC	Satélite Geoestacionário de Defesa e Comunicações (Geostationary Satellite of Defense and Communications)
SGSO	Sistema de Gerenciamento de Segurança Operacional (Operational Security Management System)
SIGMA	Sistema Integrado de Gerenciamento de Movimento Aéreo (Integrated Air Movement and Management System)
SIMIS	Seeker Image and Missile Simulation
SIPAM	Sistema Integrado de Proteção da Amazônia (Amazon Protection System)
SISCEAB	Sistema de Controle do Espaço Aéreo Brasileiro (Airspace Control System)
SISDABRA	Sistema de Defesa Aeroespacial Brasileiro (Airspace Defense System)
SISDACTA	Sistema de Defesa Aérea e Controle de Tráfego Aéreo (Integrated System of Air Defense and Air Traffic Control)
SISDCAER	Sistema de Defesa Cibernética (Cybernetic Defense System)
SIVAM	Sistema de Vigilância da Amazônia (Amazon Surveillance System)
SMDC	Sistema Militar de Defesa Cibernética (Military Cyber Defense System)
SME	small and medium enterprises
SOFA	Status of Forces Agreement
SPS	self-protection system

SRAAM	short-range air-to-air missile
SRBC	Simulador de Radar de Baixo Custo (Baixo Custo Radar simulator)
SRPA	small remotely piloted aircraft
SSA	Secretariat of Strategic Affairs
SUS	Sistema Único de Saúde (Unified Health System)
SWIR	short wavelength infrared
TCC	troop contributing countries
TMA	terminal maneuvering area/terminal control area
TMCI	The Military Conflict Institute
ToT	transfer of technology
UA	unmanned aircraft
UN	United Nations
UNDP	United Nations Development Program
UNIFA	Universidade da Força Aérea (Air Force University)
UNIFIL	United Nations Interim Force in Lebanon
UNODC	United Nations Office on Drugs and Crime
UNPCRS	United Nations Peacekeeping Capability Readiness System
UNSAS	United Nations Standby Arrangement System
UNSCOB	United Nations Special Committee on the Balkans
US, USA	United States of America
USAF	United States Air Force
USSF	United States Space Force
USSR	Union of Soviet Socialist Republics
WHO	World Health Organization

Bibliography

Abbott, Daniel, ed. *The Handbook of Fifth-Generation Warfare (5GW)*. Ann Arbor, MI: Nimble Books LLC, 2010.

Adam, Lénaïck, Gabriel Serville, Bruno Duvergé, Stéphanie Kerbarh, Josette Manin, Gérard Menuel, Annie Chapelier, et al. *RAPPORT FAIT AU NOM DE LA COMMISSION D'ENQUÊTE sur la lutte contre l'orpaillage illégal en Guyane* [REPORT DONE ON BEHALF OF THE COMMISSION OF INQUIRY into the fight against illegal gold panning in Guyana]. National Assembly Constitution of October 4, 1958, Fifteenth Legislature. Registered at the Presidency of the National Assembly on July 21, 2021. https://www.bio-plateaux.org/.

Adey, Peter. *Aerial Life: Spaces, Mobilities, Affects*. 1st ed. Malden, MA: Wiley-Blackwell, 2010.

———. "Aeromobilities: Geographies, Subjects and Vision." *Geography Compass* 2, no. 5 (2008): 1318–36. https://doi.org/10.1111/j.1749-8198.2008.00149.x.

AEL Sistemas. "Sistemas para o KC-390" [KC-390 systems]. Accessed September 20, 2017. https://www.ael.com.br/.

Affonso, José Augusto Crepaldi. "A POLÍTICA DE OFFSET DA AERONÁUTICA NO ÂMBITO DA ESTRATÉGIA NACIONAL DE DEFESA" [The offset policy of the aeronautics in the framework of the national defense strategy]. Concurso de Artigos sobre o Livro Branco de Defesa Nacional, Brasília, DF, 2012.

Águas, Carla Ladeira Pimentel. "A tripla face da fronteira: reflexões sobre o dinamismo das relações fronteiriças a partir de três modelos de análise" [The triple face of the border: Reflections on the dynamism of border relations from three analysis models]. *Forum Sociológico*. Série II, no. 23 (November 1, 2013). https://doi.org/10.4000/sociologico.842.

Aguilar, Sergio Luiz Cruz. "A Participação Do Brasil Nas Operações de Paz: Passado, Presente e Futuro" [Brazil's participation in peacekeeping operations: Past, present and future]. *Brasiliana: Journal for Brazilian Studies* 3, no. 2 (March 24, 2015): 113–41.

———. "ATLÂNTICO SUL: AS RELAÇÕES DO BRASIL COM OS PAÍSES AFRICANOS NO CAMPO DA SEGURANÇA E DEFESA" [South Atlantic: The relations between Brazil and Africa in the field of security and defense]. *AUSTRAL: Brazilian Journal of*

Strategy & International Relations 2, no. 4 (September 11, 2013). https://doi.org/10.22456/2238-6912.41288.

Air Force Doctrine Publication (AFDP) 4-0. *Combat Support.* January 4, 2020. https://www.doctrine.af.mil/.

Airports Council International. "Annual World Airport Traffic Report, 2019," March 23, 2020. https://store.aci.aero/.

Ahlgren, Linnea. "The Future of Fuel Source Sustainability with Airlines." Simpleflying.com, June 4, 2020. Accessed June 11, 2020. https://simpleflying.com/.

Albert, Bruce. "Terras indígenas, política ambiental e geopolítica militar no desenvolvimento da Amazônia: a propósito do caso Yanomami" [Indigenous lands, environmental policy and military geopolitics in the development of the Amazon: The purpose of the Yanomami case]. In *Amazônia: a fronteira agrícola 20 anos depois,* edited by Philippe Léna and Aldélia Engrácia de Oliveira, 37–58. Bélem, Brazil: Editora CEJUP, 1992.

Almeida, Carlos Wellington Leite de. "Política de defesa no Brasil: considerações do ponto de vista das políticas públicas" [Defense policy in Brazil: Considerations from the point of view of public policies]. *Opinião Pública* 16 (June 2010): 220–50. https://periodicos.sbu.unicamp.br/.

———. "Sistema de vigilância da Amazônia – SIVAM, Perspectivas da economia de defesa" [Amazon Surveillance System – SIVAM, Defense Economy Outlook]. *A Defesa Nacional* 88, no. 793 (2002): 42–57. http://www.ebrevistas.eb.mil.br/.

Al-Rodhan, Nayef R. F. *Meta-Geopolitics of Outer Space: An Analysis of Space Power, Security and Governance.* Houndmills, UK: Palgrave Macmillan, 2012.

Alvim, Mariana. "How France Preserves and Explores Its Piece of the Amazon in French Guiana." BBC News Brazil, September 1, 2019. https://www.bbc.com/.

Army Doctrine Publication (ADP) 4-0. *Sustainment.* July 2019. https://armypubs.army.mil/.

Art, Robert J., and Kenneth N. Waltz. *The Use of Force: Military Power and International Politics.* 4th ed. Lanham, MD: Rowman & Littlefield Publishers, 1993.

Assis, Stephan Delgado. "INTERDEPENDÊNCIA ENTRE SUBSIDIÁRIAS ESTRANGEIRAS E ADAPTAÇÃO DE MULTINACIONAIS: simulações a partir do modelo NK" [Interdependence between foreign subsidiaries and adaptation of multinationals:

Simulations based on the NK model]. Master's thesis, Pontific Catholic University of Minas Gerais, 2015. http://www.biblioteca.pucminas.br/.

Australian Army. Land Warfare Doctrine 4-0. *Logistics 2018. Australian Government, 2018.* https://australianarmycadets.files.wordpress.com/.

Aviation Transportation Action Group (ATAG). "Aviation: Benefits Beyond Borders." September 30, 2020. https://aviationbenefits.org/.

Aydin, Ilayda. *Geopolitics of Outer Space: Global Security and Development.* Washington, DC: Westphalia Press, 2019.

Bach, L., Patrick Cohendet, and Eric Schenk. "Technological Transfers from the European Space Programs: A Dynamic View and Comparison with Other R&D Projects." *Journal of Technology Transfer* 27 (January 1, 2003): 321–38. https://doi.org/10.1023/A:1020259522902.

Baptista Júnior, Carlos de Almeida. "Ministério da Defesa: Realidade, Desafios e Perspectivas" [Ministry of Defense: Reality, challenges and perspectives]. Course paper, Escola Superior de Guerra, 2008.

Barretto, Vicente de Paulo. "Bioética, biodireito e direitos humanos" [Bioethics, biolaw and human rights]. In *Teoria dos Direitos Fundamentais*, 2nd ed. Rio de Janeiro: Renovar, 2001.

Barroso, Luís Roberto. "Novo Direito Constitucional Brasileiro (Pós-modernidade, teoria crítica e pós-positivismo)" [New Brazilian Constitutional Law (Post-modernity, critical theory and post-positivism)]. In *A nova interpretação constitucional: ponderação, direitos fundamentais e relações privadas*, 1st ed. Rio de Janeiro: Renovar, 2003.

Bartles, Charles K. "Getting Gerasimov Right." *Military Review* 96, no. 1 (January–February 2016): 30–38. https://www.armyupress.army.mil/.

Basseto, Murilo. "Com radar, satélite e ataque a pistas, veja como a FAB defende e controla os céus da Amazônia" [With radar, satellite and runway attack, see how the FAB defends and controls the skies of the Amazon]. Aeroin, August 6, 2022. https://aeroin.net/.

Becker, Bertha K. "A Amazônia e a política ambiental brasileira" [The Amazon and Brazilian environmental policy]. In *Território, territórios: Ensaios sobre o ordenamento territorial*, 22–40. Rio de Janeiro: Lamparina, 2007.

Bertalanffy, Ludwig von. *Teoria geral dos sistemas: fundamentos, desenvolvimento e aplicações* [General systems theory: Fundamentals, development and applications]. Petropolis, Brazil: Vozes, 2010.

Bhabha, Homi K. *O Local Da Cultura* [The place of culture]. Belo Horizonte, Brazil: UFMG, 1998. http://site.livrariacultura.com.br/.

Bittencourt, Armando de Senna. "A presença da Marinha do Brasil em missão pioneira de manutenção de paz – a comissão especial da ONU nos Bálcãs (UNSCOB), 1948–1951" [The presence of the Brazilian Navy in a pioneering peacekeeping mission: The UN special commission in the Balkans, (UNSCOB), 1948–1951]. *Revista Navigator* 5, no. 10 (2009): 85–92. http://portaldeperiodicos.marinha.mil.br/.

Bobbio, Norberto. *A era dos direitos* [The era of rights]. Translated by Carlos Nelson Coutinho. Rio de Janeiro: Elsevier, 2004.

———. *O positivismo jurídico: lições de filosofia do direito* [Legal positivism: Lessons in the philosophy of law]. Translated by Márcio Pugliesi, Edson Bini, and Carlos E. Rodrigues. São Paulo: Ícone, 1995.

Bohrer, Clóvis A. *Eduardo Pacheco e Chaves: Pioneiro e Ás Da Aviação Brasileira* [Eduardo Pacheco Chaves: Pioneer and ace of Brazilian aviation]. Rio de Janeiro: Instituto Histórico-Cultural da Aeronáutica, 2014. https://www2.fab.mil.br/.

Bonnett, Alastair. *What Is Geography?* London: Sage Publications, 2008.

Boyne, Walter J. *The Influence of Air Power Upon History.* Barnsley, UK: Pen and Sword Aviation, 2005.

Bozeman, Barry. "Technology Transfer and Public Policy: A Review of Research and Theory." Research Policy 29, no. 4 (April 1, 2000): 627–55. https://doi.org/10.1016/S0048-7333(99)00093-1.

Bozeman, Barry, Heather Rimes, and Jan Youtie. "The Evolving State-of-the-Art in Technology Transfer Research: Revisiting the Contingent Effectiveness Model." *Research Policy* 44, no. 1 (February 1, 2015): 34–49. https://doi.org/10.1016/j.respol.2014.06.008.

Bracey, Djuan. "O Brasil e as operações de manutenção da paz da ONU: os casos do Timor Leste e Haiti" [Brazil and UN peacekeeping operations: The cases of East Timor and Haiti]. *Contexto Internacional* 33, no. 2 (December 2011): 315–31. https://doi.org/10.1590/S0102-85292011000200003.

Braga, Carlos Chagas Vianna. "MINUSTAH and the Security Environment in Haiti: Brazil and South American Cooperation in the Field." *International Peacekeeping* 17, no. 5 (November 1, 2010): 711–22. https://doi.org/10.1080/13533312.2010.516979.

Buck, Susan J. *The Global Commons: An Introduction.* Washington, DC: Island Press, 1998.

Budiansky, Stephen. *Air Power: The Men, Machines, and Ideas That Revolutionized War, from Kitty Hawk to Iraq.* New York: Penguin Books, 2004.
Builder, Carl. *The Masks of War: American Military Styles in Strategy and Analysis, a RAND Corporation Research Study.* Baltimore: Johns Hopkins University Press, 1989.
Bunge, Mario. *Scientific Materialism.* Dordrecht, Hollan: Springer Netherlands, 2011.
———. *Treatise on Basic Philosophy: Semantics I: Sense and Reference:* 1. Softcover reprint of the original 1st 1974 ed. Dordrecht: Springer, 1974.
———. "Why Axiomatize?" *Foundations of Science* 22, no. 4 (December 1, 2017): 695–707. https://doi.org/10.1007/s10699-016-9493-8.
Buss, Samuel R., ed. *Handbook of Proof Theory.* Vol. 137 of *Studies in Logic and the Foundations of Mathematics*, edited by S. Abramsky, S. Artemov, R. A. Shore, and A. S. Troelstra. Amsterdam: Elsevier, 1998.
Caggiano, Giovanni, Efrem Castelnuovo, and Richard Kima. "The Global Effects of Covid-19-Induced Uncertainty." *Economics Letters* 194 (September 2020): 109–392. https://doi.org/10.1016/j.econlet.2020.109392.
Calaza, Claudio Passos. "Inteligência cultural: novos parâmetros na formação do oficial ante a nova geração de conflitos" [Cultural intelligence: New parameters in the formation of the officer in the face of the new generation of conflicts]. Presented at the V Encontro Pedagógico do Ensino Superior Militar, Rio de Janeiro, Escola Naval, September 2–6, 2012. http://www.redebim.dphdm.mar.mil.br/.
Cannabrava, Ivan Oliveira. "O Brasil e as operações de manutenção de paz" [Brazil and peacekeeping operations]. *Revista Política Externa* 5, no. 3 (December 1996): 93–105.
Carnaúba, Valquíria, and Ana Cristina Coccolo. "Um Desafio Do Século XXI" [A 21st century challenge]. *Entreteses*, June 2016.
Carr, E. H. *Vinte Anos de Crise. 1919–1939. Uma Introdução ao Estudo das Relações Internacionais* [Twenty years of crisis, 1919–1939: An introduction to the study of international relations]. 2nd ed. São Paulo: Imprensa Oficial do Estado de São Paulo, 2001.
Castro, Thales. *Teoria das relações internacionais* [International relations theory]. 2nd ed. Brasília: Fundação Alexandre de Gusmão, 2016.
Cavalcanti, Agostinho P. B., and Adler G. Viadana. "Fundamentos Históricos da Geografia: Contribuições do Pensamento Filosóf-

ico na Grécia Antiga" [Historical foundations of geography: Contributions of philosophical thought in ancient Greece]. In *História do pensamento geográfico e epistemologia em Geografia*, edited by Paulo R. Teixeira de Godoy, 11–34. São Paulo: Editora UNESP, 2010. http://books.scielo.org/.

Central Intelligence Agency (CIA). "Tonga." The World Factbook. Accessed March 25, 2020. https://web.archive.org/.

Chun, Clayton. *Aerospace Power in the Twenty-First Century: A Basic Primer.* Scotts Valley, CA: CreateSpace Independent Publishing Platform, 2012.

Cinelli, Carlos Frederico. *Direito Internacional Humanitário - Ética e Legitimidade no Uso da Força em Conflitos Armados* [International humanitarian law: Ethics and legitimacy in the use of force in armed conflicts]. 2nd ed. Edited by José Ernani de Carvalho Pacheco. Curitiba, Brazil: Juruá Editora, 2016.

Ciocan, Florian. "Perspectives on Interoperability Integration within NATO Defense Planning Process." *Journal of Defense Resources Management* 2, no. 2 (2011): 53–66. http://www.jodrm.eu/.

Clausewitz, Carl von. *Da Guerra* [On War]. Brasília: Editora UnB, 1979.

Cohen, Raphael S., Nathan Chandler, Shira Efron, Bryan Frederick, Eugeniu Han, Kurt Klein, Forrest E. Morgan, Ashley L. Rhoades, Howard J. Shatz, and Yuliya Shokh. "Peering into the Crystal Ball: Holistically Assessing the Future of Warfare." RAND Corporation, May 11, 2020. https://www.rand.org/.

Collins, John M. *Military Geography: For Professionals and the Public.* 1st ed. Washington, DC: National Defense University Press, 1998.

Comte, Auguste. *Curso de Filosofia Positiva* [Positive philosophy course]. São Paulo: Abril Cultural, 1983.

"Constitución española de 1978" [The Spanish constitution of 1978]. Chamber of Congreso de los Diputados, 2003. https://app.congreso.es/.

Coordinating Committee for the Implementation of SIVAM Project. "O Sistema de Vigilância Da Amazônia (SIVAM)" [The Amazon surveillance system (SIVAM)]. Brasília: Brazilian Ministry of Aeronautics, May 31, 2005.

COPUOS (Committee on the Peaceful Uses of Outer Space). Report of the Committee on the Peaceful Uses of Outer Space (Official Records, Fifty-sixth Session, Supplement No. 20 [A/56/20]). New York: General Assembly of the Committee on the Peaceful Uses of Outer Space, 2001.

Correa, Gilberto Mohr. "Resultados Da Política de Offset Da Aeronáutica: Incremento Nas Capacidades Tecnológicas Das Organizações Do Setor Aeroespacial Brasileiro" [Results of the aeronautics offset policy: Increase in the technological capabilities of organizations in the Brazilian aerospace sector]. Master's thesis, Instituto Tecnológico de Aeronáutica, 2017. http://www.bdita.bibl.ita.br/.

Corte Interamericana de Direitos Humanos. Caso Gomes Lund e Outros ("Guerrilha do Araguaia") v. Brasil, SENTENÇA DE 24 DE NOVEMBRO DE 2010 [Case of Gomes Lund and Others ("Guerrilla of Araguaia") v. Brazil, Judgment of November 24, 2010]. https://www.corteidh.or.cr/.

Cosgrove, Denis. "Contested Global Visions: *One-World, Whole-Earth, and the Apollo Space Photographs.*" *Annals of the Association of American Geographers* 84, no. 2 (June 1, 1994): 270–94. https://doi.org/10.1111/j.1467-8306.1994.tb01738.x.

Costa, Marcos Phelipe Dias da. "Interoperabilidade: o impacto da diminuição das ações com tropas e meios nas operações de adestramento conjunto" [Interoperability: The impact of decreasing troop and asset actions on joint training operations]. Thesis, Escola Superior de Guerra, 2018. https://repositorio.esg.br/handle/123456789/909.

Creveld, Martin van. *The Rise and Decline of the State.* Cambridge: Cambridge University Press, 1999. https://doi.org/10.1017/CBO9780511497599.

Croda, Julio, Wanderson Kleber de Oliveira, Rodrigo Lins Frutuoso, Luiz Henrique Mandetta, Djane Clarys Baia-da-Silva, José Diego Brito-Sousa, Wuelton Marcelo Monteiro, and Marcus Vinícius Guimarães Lacerda. "COVID-19 in Brazil: Advantages of a Socialized Unified Health System and Preparation to Contain Cases." *Journal of the Brazilian Society of Tropical Medicine* 53 (April 2020). https://doi.org/10.1590/0037-8682-0167-2020.

Cwerner, Saulo, Sven Kesselring, and John Urry, eds. *Aeromobilities.* London: Routledge, 2009. https://doi.org/10.4324/9780203930564.

Cysne, Maria do Rosário de Fátima Portela. "Transferência de tecnologia entre a universidade e a indústria" [Technology transfer between university and industry]. *Encontros Bibli: revista eletrônica de biblioteconomia e ciência da informação* 10, no. 20 (January 1, 2005): 54–74. https://doi.org/10.5007/1518-2924.2005v10n20p54.

"Declaration of the First Meeting of Equatorial Countries" (signed in Bogota December 3, 1976, by heads of delegations from Brazil, Colombia, Congo, Ecuador, Indonesia, Kenya, Uganda, and Zaire). *Space Law* (database). Accessed July 31, 2020. https://www.jaxa.jp/.

Deen, Thalif. "África Do Sul Continua Sendo a Maior Fábrica de Armas Do Sul" [South Africa remains the largest weapons factory in the south]. Inter Press Service Noticias, December 13, 2013.

Defense Industry Daily. "South Africa, Brazil's A-Darter SRAAM Hits Target." Defense Industry Daily, October 4, 2019. http://www.defenseindustrydaily.com/.

Defesanet. "Poder Executivo Entrega Atualizações Da PND, END e LBDN Ao Congresso Nacional" [Executive Branch delivers PND, END and LBDN updates to the National Congress]. Defesanet. com, July 30, 2020. https://www.defesanet.com.br/.

Depaula, Pablo Domingo. "Predictores Globales de la Performance de Estudiantes Militares" [Global predictors of military students' performance]. *Ciencias Psicológicas* 6, no. 2 (2012): 135–48. https://www.redalyc.org/.

De Seversky, Alexander P. *Air Power: Key to Survival.* 1st ed. New York: Simon and Schuster, 1950.

Deudney, Daniel. *Space: The High Frontier in Perspective.* Washington, DC: Worldwatch Institute, 1982.

Doboš, Bohumil. *Geopolitics of the Outer Space: A European Perspective.* 1st ed. New York: Springer, 2018.

Dolman, Everett C. *Astropolitik: Classical Geopolitics in the Space Age.* 1st ed. London: Routledge, 2002.

Dorn, A. Walter. *Air Power in UN Operations: Wings for Peace.* London: Routledge, 2016. https://doi.org/10.4324/9781315566313.

Douhet, Giulio. *O Domínio Do Ar* [Command of the air]. Rio de Janeiro: Instituto Histórico da Aeronáutica, 1988.

DuBois, Edmund L., Wayne P. Hughes Jr., and Lawrence J. Low. *A Concise Theory of Combat.* Institute for Joint Warfare Analysis, Naval Postgraduate School technical report, NPS-IJWA-97-001. All Technical Reports Collection. Monterey, CA: Institute for Joint Warfare Analysis, 1997. https://calhoun.nps.edu/.

Earley, P. Christopher, and Elaine Mosakowski. "Cultural Intelligence." *Harvard Business Review,* October 2004. https://hbr.org/.

Elden, Stuart. "Secure the Volume: Vertical Geopolitics and the Depth of Power." *Political Geography* 34 (May 2013): 35–51. https://doi.org/10.1016/j.polgeo.2012.12.009.

Embraer. *Manual de Voo - Manual Suplementar de Dados de Desempenho Aeronave A-29A (Monoposto) A-29B (Biposto)* [Flight

manual: Supplementary performance data manual aircraft A-29A (Single-seat) A-29B (Bi-seat)]. São Paulo: Embraer, 2007.
Esri. "What Is Geoprocessing?" Esri.com, 2020. https://pro.arcgis.com/.
European Space Agency. "Space Debris by the Numbers." European Space Agency. Accessed May 8, 2020. https://www.esa.int/.
Ferrajoli, Luigi. *A Soberania No Mundo Moderno* [Sovereignty in the modern world]. 1st ed. São Paulo: Martins Fontes, 2003.
Ferreira, Luciano Vaz. *Direito Internacional da Guerra* [International law of war]. 1st ed. Jundiaí, Brazil: Paco Editorial, 2014.
Ferreira, Raphael Ignácio. "Mudanças Na Política de Defesa Cibernética Brasileira Após o Escândalo de Espionagem Norteamericano" [Changes in Brazilian cyber defense policy after the North American espionage scandal]. In *Anais Do ERABED*. São Paulo: ERABED, 2019. https://www.erabedsudeste2019.abedef.org/.
Flemes, Daniel. "O Brasil na iniciativa BRIC: soft balancing numa ordem global em mudança?" [Brazil in the BRIC initiative: Soft balancing in the shifting world order?]. *Revista Brasileira de Política Internacional* 53 (July 2010): 141–56. https://doi.org/10.1590/S0034-73292010000100008.
Franchitto, M., and C. Rebouças. *Relatório Sobre Tecnologias Do Programa A-Darter* [A-Darter program technologies report]. Technical report. Pretória:[s.n.], 2009.
Freire, Laura Maria Corrêa de Sá. "REGIÃO AMAZÔNICA: Novas Ameaças e Possíveis Respostas" [Amazon region: New threats and possible answers]. *Revista da Escola Superior de Guerra*, no. 44 (2004): 71–127. https://doi.org/10.47240/revistadaesg.v20i44.347.
Freire, Maria Eduarda Laryssa Silva, and Mariana Pimenta Oliveira Baccarini. "Uma análise neo-institucional da adoção de jointness pelos Estados Unidos" [A neo-institutional analysis of the United States' adoption of jointness]. *Revista Brasileira de Estudos de Defesa* 7, no. 1 (December 29, 2020). https://doi.org/10.26792/rbed.v7n1.2020.75181.
Frenken, Koen. "A Complexity Approach to Innovation Networks: The Case of the Aircraft Industry (1909–1997)." *Research Policy 29*, no. 2 (2000): 257–72. https://doi.org/10.1016/S0048-7333(99)00064-5.
———. "Modelling the Organisation of Innovative Activity Using the NK-Model." Paper prepared for the Nelson-and-Winter Conference, Aalborg, June 12–16, 2001. https://citeseerx.ist.psu.edu/.
Fukuyama, Francis. *O fim da história e o último homem* [The end of history and the last man]. Rio de Janeiro: Biblioteca do Exército, 1998.

Gavetti, Giovanni, and Daniel Levinthal. "Looking Forward and Looking Backward: Cognitive and Experiential Search." *Administrative Science Quarterly* 45, no. 1 (March 1, 2000): 113–37. https://doi.org/10.2307/2666981.

Ginzburg, Carlo. *O queijo e os vermes: O cotidiano e as ideias de um moleiro perseguido pela Inquisição* [The cheese and the worms: The daily life and ideas of a miller persecuted by the Inquisition]. Translated by Maria Betânia Amoroso. 1st ed. São Paulo: Companhia das Letras, 1987.

Global Fire Power. "2021 Military Strength Ranking." GlobalFirePower.com. Accessed September 15, 2022. https://www.globalfirepower.com/.

Góes, Guilherme Sandoval. "GEOPOLÍTICA E CONSTITUIÇÃO À LUZ DO ESTADO DEMOCRÁTICO DE DIREITO" [Geopolitics and the constitution in light of the democratic constitutional state]. *AUSTRAL: Brazilian Journal of Strategy & International Relations* 9, no. 18 (2021). https://doi.org/10.22456/2238-6912.108981.

———. "O Geodireito e os centros mundiais de poder." Apresentação realizada no VII Encontro Nacional de Estudos Estratégicos, 06 a 08 de novembro de 2007, Gabinete de Segurança Institucional da Presidência da República, Brasília/DF.

Goldoni, Luiz Rogério Franco, and Sandro De Nazareth Ciribelli. "RELAÇÕES DO BRASIL COM ÁFRICA DO SUL E ANGOLA: ESFORÇOS PARA A MANUTENÇÃO DA SEGURANÇA NO ATLÂNTICO SUL" [Relations of Brazil with South Africa and Angola: Efforts for the maintenance of security in the South Atlantic]. *AUSTRAL: Brazilian Journal of Strategy & International Relations* 5, no. 9 (October 27, 2016). https://doi.org/10.22456/2238-6912.63851.

Gonçalves, Joanisval Brito. *TRIBUNAL DE NUREMBERG 1945–1946: A GÊNESE DE UMA NOVA ORDEM NO DIREITO INTERNACIONAL* [Nuremberg court 1945–1946: The genesis of a new order in international law]. 1st ed. Rio de Janeiro: Renovar, 2004.

Gonçalves, Nelson Augusto Bacellar. "A Força Aérea Brasileira na Missão de Estabilização das Nações Unidas no Haiti: A Dependência de uma Aeronave de Transporte Estratégica" [The Brazilian Air Force in the United Nations stabilization mission in Haiti: Dependence on a strategic transport aircraft]. Master's thesis, UNIFA, 2016. https://www2.fab.mil.br/.

Gray, Colin S. "Inescapable Geography." In *Geopolitics, Geography and Strategy*, edited by Colin S. Gray and Geoffrey Sloan. 16–77. Portland, OR: Frank Cass, 1999.

Guimarães, Álvaro Wolnei. "DECEA Inicia Implantação Da Plataforma X-4000 No Paraguai" [DECEA starts deployment of the X-4000 platform in Paraguay]. Brazilian Air Force, Department of Airspace Control. October 10, 2018. https://www.decea.mil.br/.

Hacisuleyman, Ezgi, Caryn Hale, Yuhki Saito, Nathalie E. Blachere, Marissa Bergh, Erin G. Conlon, Dennis J. Schaefer-Babajew, et al. "Vaccine Breakthrough Infections with SARS-CoV-2 Variants." *The New England Journal of Medicine* 384, no. 23 (June 10, 2021): 2212–18. https://doi.org/10.1056/NEJMoa2105000.

Hale, Charles R. "Cultural Politics of Identity in Latin America." *Annual Review of Anthropology* 26 (1997): 56790. https://www.jstor.org/stable/2952535.

Harrison, Todd. *International Perspectives on Space Weapons. CSIS Aerospace Security Project.* Washington, DC: Center for Strategic and International Studies, May 2020. https://aerospace.csis.org/.

Hartmann, Kim, and Keir Giles. "UAV Exploitation: A New Domain for Cyber Power." In *2016 8th International Conference on Cyber Conflict (CyCon)*, edited by N. Pissanidis, H. Rõigas, M. Veenendaal, 205–21. Brussels, NATO, 2016. https://doi.org/10.1109/CYCON.2016.7529436.

Hartshorne, Richard. *Perspective on the Nature of Geography.* Chicago: Association of American Geographers, 1959.

Hayward, Justin. "Airline Alliances: What Are They & What Are the Benefits?" *Simple Flying,* June 24, 2020. https://simpleflying.com/.

Heye, Thomas Ferdinand. "Democracia, controle civil e gastos militares no Pós-Guerra Fria: uma análise realista" [Democracy, civil control and post–Cold War military spending: A realistic analysis]. *Carta Internacional* 10, no. 1 (April 15, 2015): 105–34. https://doi.org/10.21530/ci.v10n1.2015.206.

Hillen, John. "Peacekeeping at the Speed of Sound: The Relevancy of Airpower Doctrine in Operations Other than War." *Airpower Journal* (Winter 1998). https://apps.dtic.mil/.

Hoffman, Frank G. *Conflict in the 21st Century: The Rise of Hybrid Wars.* Arlington, VA: Potomac Institute for Policy Studies, 2007. https://www.potomacinstitute.org/.

Holland, John H. *Hidden Order: How Adaptation Builds Complexity.* Reading, MA: Helix Books, 1995.

Holt-Jensen, Arild. *Geography: History and Concepts. A Student's Guide.* 4th ed. Thousand Oaks, CA: Sage Publications, 2009.
Hugget, Richard John, and Mike Robinson. "Introduction." In *Companion Encyclopedia of Geography: The Environment and Humankind*, edited by Ian Douglas, Richard John Hugget, and Mike Robinson. London: Routledge, 1997. https://doi.org/10.4324/9780203416822.
Huntington, Samuel P. *The Clash of Civilizations and the Remaking of World Order.* New York: Simon & Schuster, 2007.
Huston, James A. *The Sinews of War: Army Logistics, 1775–1953.* Washington, DC: Department of Defense Center of Military History, 1997.
Ignacio, Julia. "ECO-92: What was the conference and what were its main results?" Politize.com, November 23, 2020. https://www.politize.com.br/.
Iha, Bruno. "FAB encerra participação no Exercício Internacional Green Flag West" [FAB ends its participation in the international exercise Green Flag West]. Brazilian Air Force (website). Accessed May 2, 2022. https://www.fab.mil.br/.
Instituto Histórico-Cultural da Aeronáutica. *História Geral da Aeronáutica Brasileira: dos Primórdios Até 1920* [General history of Brazilian aeronautics: From the beginnings to 1920]. Vol. 1. Rio de Janeiro: Itatiaia, 1988.
International Air Transport Association. *Annual Review 2019.* 75th Annual General Meeting. Seoul: International Air Transport Association, June 2019. https://www.iata.org/.
International Civil Aviation Organization (ICAO). *2016–2030 Global Air Navigation Plan.* Doc 9750-AN/963. 5th ed. Montreal: ICAO, 2016. https://www.icao.int/.
———. *Caribbean/South American Air Traffic Flow Management Concept of Operation (CAR/SAM CONOPS ATFM).* Mexico City: International Civil Aviation Organization, October 2006. https://www.icao.int/.
———. "CAR/SAM Seminar in Preparation of Eleventh Air Navigation Conference (AN-Conf/11)." Montreal: International Civil Aviation Organization, 2003. September 22, 2003. https://www.icao.int/.
———. *Convention on International Civil Aviation, Doc. 7300/9.* 9th ed. Montreal: International Civil Aviation Organization, 2006. https://www.icao.int/.

———. *Global Air Traffic Management Operational Concept, Doc. 9854*. 1st ed. Montreal: International Civil Aviation Organization, 2005.

———. *Global Air Transport Outlook to 2030 and Trends to 2040*. Montreal: International Civil Aviation Organization, 2013.

———. *Proyecto Regional RLA/06/901*. Lima, Perú: International Civil Aviation Organization, 2018. https://1library.co/.

International Committee of the Red Cross (ICRC). "A Guide to the Legal Review of New Weapons, Means and Methods of Warfare: Measures to Implement Article 36 of Additional Protocol I of 1977: International Committee of the Red Cross Geneva, January 2006." *International Review of the Red Cross* 88, no. 864 (December 2006): 931–56. https://www.icrc.org/

———. "Como o Direito Internacional Humanitário define 'conflitos armados'?" [How does international humanitarian law define "armed conflicts"?]. ICRC, March 2008. https://www.icrc.org/.

———. "Geneva Conventions of 1949, Additional Protocols and Their Commentaries." ICRC. https://ihl-databases.icrc.org/.

International Monetary Fund. "Policy Responses to COVID-19." IMF.org, July 2, 2021. https://www.imf.org/.

Ispas, Lucian, and Paul Tudorache. "Cultural Interoperability—Prerequisite for the Response to Hybrid Threats." In *International Conference KNOWLEDGE-BASED ORGANIZATION 23*, no. 1 (June 2017): 161–65. https://doi.org/10.1515/kbo-2017-0025.

Jameson, Fredric. *Pós-Modernismo - A Lógica Cultural Do Capitalismo Tardio* [Postmodernism: The cultural logic of late capitalism]. 2nd ed. São Paulo: Editora Ática, 2002.

Jayme, Jonathan. "FAB intercepta aeronave carregada com 165 kg de drogas no Amazonas" [FAB intercepts aircraft loaded with 165 kg of drugs in Amazonas]. Brazilian air force (website), February 19, 2022. https://www.fab.mil.br/.

———. "Míssil A-Darter Conclui Testes Na África Do Sul" [A-Darter Missile Completes Testing in South Africa]. Brazilian air force (website), September 28, 2018. https://www.fab.mil.br/.

John, David F. "Unmanned Systems in Perspective." Master's thesis, School of Advanced Military Studies, 2014. https://apps.dtic.mil/.

Johnson-Freese, Joan. *Space as a Strategic Asset*. New York: Columbia University Press, 2007.

Kant, Immanuel. *Perpetual Peace*. Translated by M. Campbell Smith. London: George Allen & Unwin Ltd., 1903.

Kauffman, Stuart A. *The Origins of Order: Self-Organization and Selection in Evolution.* New York: Oxford University Press, 1993.

Kenkel, Kai Michael. "Brazil's Peacekeeping and Peacebuilding Policies in Africa." *Journal of International Peacekeeping* 17, no. 3–4 (January 1, 2013): 272–92. https://doi.org/10.1163/18754112-1704006.

Kitchin, Rob, and Nigel Thrift, eds. *International Encyclopedia of Human Geography.* Vol. 1. Amsterdam: Elsevier, 2009. https://www.sciencedirect.com/.

Kjellén, Rudolf. *Staten som lifsform* [The state as a form of life]. Politiska Handböcker III. Stockholm: Hugo Gebers Förlag, 1916.

Klein, John J. *Space Warfare: Strategy, Principles and Policy.* New York: Routledge, 2012.

Korybko, Andrew. *The Law of Hybrid War: Eastern Hemisphere.* Self-published, Kindle, 2017.

Kress, Moshe. *Operational Logistics: The Art and Science of Sustaining Military Operations.* 2nd ed. New York: Springer, 2015.

Kundu, Nirmal, Chandan Bhar, and Visvesvaran Pandurangan. "Development of Framework for an Integrated Model for Technology Transfer." *Indian Journal of Science and Technology* 8, no. 35 (December 18, 2015): 1–14. https://doi.org/10.17485/ijst/2015/v8i35/74280.

Lavenère-Wanderley, Nelson Freire. *Historia Da Força Aerea Brasileira* [History of the Brazilian Air Force]. Brasília: Brazilian Air Force Ministry, 1957.

Lemos Júnior, Francisco das Chagas. "Jointness, pensamento conjunto e conjuntez : estudo comparativo entre o processo de reforma do exército estadunidense (1973–1991) e a modernizacão da estrutura militar brasileira (1999–2020)" [Jointness, joint thinking and conjuncture: a comparative study between the reform process of the US army (1973–1991) and the modernization of the Brazilian military structure (1999–2020)]. Course paper, Escola Superior de Guerra, 2020.

Lewis, Bernard. "The Roots of Muslim Rage." *The Atlantic,* September 1, 1990. https://www.theatlantic.com/.

Liang, Qiao, and Wang Xiangui. *Unrestricted Warfare: China's Master Plan to Destroy America.* Lambertville, NJ: Shadow Lawn Press, 2017.

Lind, William S. "Understanding Fourth Generation War." *Military Review* 84, no. 5 (September–October 2004): 12–16.

Lindsay, Jon R., and Erik Gartzke. "Politics by Many Other Means: The Comparative Strategic Advantages of Operational Domains."

Journal of Strategic Studies (June 1, 2020): 1–34. https://doi.org/10.1080/01402390.2020.1768372.

Lipovetsky, Gilles. *Os Tempos Hipermodernos* [Hypermodern times]. 1st ed. São Paulo: Barcarolla, 2004.

Lôbo Júnior, Joaquim Tavares. "O gerenciamento de fluxo de tráfego aéreo sob uma perspectiva geopolítica do Poder Aeroespacial brasileiro" [The management of air traffic flow from a geopolitical perspective of the Brazilian Aerospace Power]. Dissertation, Universidade da Força Aérea, 2020.

Longo, Waldimir Pirró. "Ciência e Tecnologia: Alguns Aspectos Teóricos" [Science and technology: Some theoretical aspects]. Manuscrito LS-19/87. Rio de Janeiro: Escola Superior de Guerra, 2004. https://web.archive.org/.

———. "Tecnologia militar: conceituação, importância e cerceamento" [Military technology: conceptualization, importance and restriction]. *Tensões Mundiais* 3, no. 5 (July/December 2007): 111–43. https://revistas.uece.br/.

Lonsdale, David J. "Information Power: Strategy, Geopolitics, and the Fifth Dimension." *Journal of Strategic Studies* 22, no. 2–3 (June 1, 1999): 137–57. https://doi.org/10.1080/01402399908437758.

Lopes, Inez. "Breves Considerações sobre os Princípios Constitucionais das Relações Internacionais" [Brief considerations on the constitutional principles of international relations]. *Consilium - Revista Eletrônica de Direito* 1, no. 3 (2009): 1–16.

Lopez, Andréia. *Negociação e a Inteligência Cultural. O Caso da Cultura Árabe* [Negotiation and cultural intelligence: The case of Arab culture]. 1st ed. São Paulo: Biblioteca24horas, 2013.

Lyotard, Jean-François. *A condição pós-moderna* [The postmodern condition]. 8th ed. Rio de Janeiro: José Olympio, 2004.

MacDonald, Fraser. "Anti-*Astropolitik*—Outer Space and the Orbit of Geography." *Progress in Human Geography* 31, no. 5 (October 1, 2007): 592–615. https://doi.org/10.1177/0309132507081492.

Machado, Lia. "Limites e Fronteiras: Da Alta Diplomacia Aos Circuitos Da Ilegalidade" [Limits and borders: From high diplomacy to circuits of illegality]. *Revista Território* 8 (January 1, 2000): 9–29.

Mackinder, H. J. "The Geographical Pivot of History (1904)." Reprint of Mackinder's January 25, 1904, speech at the Royal Geographical Society, in *The Geographical Journal* 170, no. 4 (2004): 298–321.

Mafra, Roberto Machado de Oliveira. *Geopolítica* [Geopolitics]. São Paulo: Sicurezza, 2006.

Magnoli, Demétrio. *História das Guerras* [The history of wars]. São Paulo: Contexto, 2006.

Mahan, A. T. *The Influence of Sea Power Upon History, 1660–1783.* 12th ed. Boston: Little, Brown and Company, 2011.

Mandarino Júnior, Raphael. *Seguranca e Defesa do Espaço Cibernético Brasileiro* [Security and defense of Brazilian cyberspace]. 1st ed. Recife, PE: Cubzac, 2010.

Mansfield, Edwin. "International Technology Transfer: Forms, Resource Requirements, and Policies." *The American Economic Review* 65, no. 2 (1975): 372–76.

Maria, Emília. "Evento Marca Encerramento Do Ciclo de Desenvolvimento Do Projeto A-Darter" [Event marks closing of the A-Darter project development cycle]. Brazilian Air Force (website), September 27, 2019. https://www.fab.mil.br/.

Marine Corps Warfighting Publication (MCWP) 3-40. *Logistic Operations,* 2017. https://www.marines.mil/.

Markoff, Marko G. "Disarmament and 'Peaceful Purposes' Provisions in the 1967 Outer Space Treaty." *Journal of Space Law* 4, no. 1 (1976): 3–22.

Marquis, Christopher G., Denton Dye, and Ross S. Kinkead. "The Advent of Jointness During the Gulf War." *Joint Force Quarterly* 85, no. 2 (2017): 76–83. https://ndupress.ndu.edu/.

Matos, Sergio Ricardo Reis, and Sheila Cristina Monteiro Matos. "Grupos Indígenas y Militares: una búsqueda sobre la variación lingüistica en lainterlocución entre estos grupos en las regiones fronterizas de la Amazonia" [Indigenous communities and military troops: A survey about the linguistic variation in the interaction between these groups in Amazon border region]. *Revista da UNIFA* 25, no. 30 (June 2012): 28–39.

Matthews, Michael R., ed. *Mario Bunge: A Centenary Festschrift.* 1st ed. New York: Springer, 2019.

Mattos, Fernando Preusser de, and Ricardo Fagundes Leães. "Relações Brasil-África do Sul: Cooperação Sul-Sul e Multilateralismo" [Brazil-South Africa Relations: South-South cooperation and multilateralism]. Seminar paper presented at the 1° Seminário Internacional de Ciência Política: Estado e Democracia em Mudança no Século XXI, Porto Alegre, Brazil, September 9–11, 2015. https://www.ufrgs.br/.

Meireles, Daisy. "Brasil Coopera Com Paraguai Na Instrução Para Controladores de Tráfego Aéreo Civis" [Brazil cooperates with

Paraguay in instruction for civil air traffic controllers]. Brazilian Air Force, Department of Airspace Control (website), November 21, 2018. https://www.decea.mil.br/.

———. "Foz Do Iguaçu Sedia a 1a Reunião Do Acordo de Cooperação Técnica Brasil-Paraguai" [Foz Do Iguaçu hosts the first meeting of the Brazil-Paraguay technical cooperation agreement]. Brazilian Air Force, Department of Airspace Control (website), August 23, 2018. https://www.decea.mil.br/.

———. "Instituto de Cartografia Aeronáutica Colabora Na Renovação de Procedimentos de Navegação Aérea Argentinos" [Institute of Aeronautical Cartography collaborates in the renewal of Argentine air navigation procedures]. Brazilian Air Force, Department of Airspace Control (website), April 11, 2017. https://www.decea.mil.br/.

Meirelles, João Filho. *O Livro de Ouro da Amazônia* [The golden book of the Amazon], 5th ed. Rio de Janeiro: Ediouro, 2009.

Mello, Celso D. de Albuquerque. *Curso de Direito Internacional Público* [Public international law course]. Vol. 1. Rio de Janeiro: Renovar, 2001.

Melzer, Nils. *International Humanitarian Law: A Comprehensive Introduction*. Geneva: International Committee of the Red Cross, 2019. https://shop.icrc.org/.

Mi, Chienkuo Michael, and Ruey-lin Chen, eds. *Naturalized Epistemology and Philosophy of Science*. Leiden, The Netherlands: Brill, 2007. https://brill.com/.

Miranda, Wando Dias, and Durbens Martins Nascimento. *Conflitos Assimétricos e o Estado: O Neoterrorismo e os Novosparadigmas para formulação de Políticas de Defesa Nacional* [Asymmetrical conflict and the state: Neoterrorism and new paradigms for the formulation of national defense policy]. In *Proceedings of the 3rd ENABRI 2011*. São Paulo, 2011. http://www.proceedings.scielo.br/.

Mitchell, William "Billy." *Winged Defense: The Development and Possibilities of Modern Air Power—Economic and Military*. With a new foreword by Robert S. Ehlers Jr. 1st ed. Tuscaloosa: Fire Ant Books Series of the University of Alabama Press, 2010.

Moltz, James Clay. *Crowded Orbits*: Conflict and Cooperation in Space. New York: Columbia University Press, 2014.

Morgenthau, Hans J. *Politics among Nations*. 6th ed. New York: Random House USA Inc., 1985.

Nardon, Laurence. "Developed Space Programmes." In *The Politics of Space: A Survey,* edited by Eligar Sadeh, 58–76. London: Routledge, 2011.

National Aeronautics and Space Administration (NASA). "Artemis." NASA.gov. October 21, 2019. https://www.nasa.gov/.

National Air and Space Intelligence Center (NASIC). "Competing in Space." NASIC Public Affairs Office, December 2018. https://media.defense.gov/.

National Intelligence Council. *Global Trends 2040: A More Contested World.* Washington, DC: Office of the Director of National Intelligence, March 2021. https://www.dni.gov/.

Naval Doctrine Publication 4. *Naval Logistics.* January 10, 1995. https://apps.dtic.mil/.

Novosseloff, Alexandra. *Keeping Peace from Above: Air Assets in UN Peace Operations.* New York: International Peace Institute, October 2017. https://www.ipinst.org/.

Nye, Joseph S., Jr. *Soft Power: The Means to Success in World Politics.* New York: Public Affairs, 2004.

———. "The Information Revolution and Power." *Current History* 113, no. 759 (2014): 19–22. https://doi.org/10.1525/curh.2014.113.759.19.

Oduntan, Gbenga. *Sovereignty and Jurisdiction in Airspace and Outer Space: Legal Criteria for Spatial Delimitation.* London: Routledge, 2012.

Oliveira, Roberto Barros de. "Ensino e interoperabilidade: um caminho para o fortalecimento da defesa nacional" [Teaching and interoperability: a path to strengthening national defense]. Course paper, Escola Superior de Guerra, 2019. https://repositorio.esg.br/.

Olivier, Darren. "A-Darter Programme Reaches Maturity." *African Defence Review,* November 9, 2018. https://www.africandefence.net/.

O'Loughlin, John, ed. *Dictionary of Geopolitics.* Westport, CT: Greenwood Press, 1994.

Omissi, David. "Technology and Repression: Air Control in Palestine 1922–36." *Journal of Strategic Studies* 13, no. 4 (December 1, 1990): 41–63. https://doi.org/10.1080/01402399008437430.

Organisation for Economic Co-operation and Development. *Geopolitical Developments and the Future of the Space Sector: OECD Futures Project on the Commercialisation of Space and the Development of Space Infrastructure: The Role of Public and Private Actors.* Paris: Organisation for Economic Co-operation and Development, May 13, 2004. https://www.oecd.org/.

Ost, François. *O Tempo do Direito* [The time of law]. Lisbon: Instituto Piaget, 2001.

Pandey, Manish, and Michael Baggs. "Why Does President Trump Want to Mine on the Moon?" BBC.com, April 12, 2020. https://www.bbc.com/.

Pascoe, David. *Airspaces*. London: Reaktion Books, 2001.

Paula, Victor Magno Gomes. "Míssil MAA-1 Piranha" [MAA-1 Piranha missile]. Juiz de Fora, Brazil: Universidade Federal de Juiz de Fora, 2009. https://docplayer.com.br/.

Pereira, Elaine Gonçalves da Costa. *O CÉU É NOSSO! A Defesa Aérea Brasileira* [The Sky Is Ours: Brazilian Air Defense]. Rio de Janeiro: The Brazilian Historical Cultural Institute of Aeronautics, 2018.

Pessoa, Tamiris Santos, and Marco Tulio Delgobbo Freitas. "A Adoção Do Modelo Joint: Reflexões Sobre Implicações No Modelo Brasileiro" [The adoption of the joint model: Reflections on implications in the Brazilian model]. *Revista Da Escola de Guerra Naval* 21, no. 2 (2016): 203–20. https://revista.egn.mar.mil.br/.

Phillips, Rhonda G. "Technology Business Incubators: How Effective as Technology Transfer Mechanisms?" *Technology in Society* 24, no. 3 (August 1, 2002): 299–316. https://doi.org/10.1016/S0160-791X(02)00010-6.

Pinheiro, Leticia Abreu. *Política externa brasileira* [Brazilian foreign policy]. 1st ed. Rio de Janeiro: Zahar, 2004.

Piovesan, Flavia. "Artigo 4o, VIII, IX e X" [Article 4, VIII, IX and X]. In *Comentários à constituição do Brasil*, edited by Gilmar Ferreira Mendes, José Gomes Canotilho, Ingo Wolfgang Sarlet, and Lenio Luiz Streck, and executive coordination of Léo Ferreira Leoncy. 2nd ed. São Paulo: Saraiva Jur, 2018.

Pires, Gustavo Calero Garriga. "A importância da Reforma 'Goldwater-Nichols' para a evolução da interoperabilidade nas Forças Armadas dos Estados Unidos da América e suas aplicações para o caso brasileiro" [The importance of the 'Goldwater-Nichols' reform for the evolution of interoperability in the armed forces of the United States of America and its applications for the Brazilian case]. Course paper, Escola Superior de Guerra, 2018. https://repositorio.esg.br/.

Porto, Newton Marcos Leone, Aline Cavalcante, Wanessa Rodrigues De Sousa, Raphael Leite, and Fernanda Pereira Garrijo. "O Meio de Transporte Aéreo como Apoio Logístico à Viabilização de Acordos Bilaterais de Goiás/Mato Grosso com Países da América do Sul" [The means of air transport as logistical support for the feasibility of bilateral agreements between Goiás/Mato Grosso with countries of South America]. *Estudos Vida e Saúde* 34, no. 7 (2007): 555–71. http://seer.pucgoias.edu.br/.

Rabkin, Jeremy, and John Yoo. *Striking Power: How Cyber, Robots, and Space Weapons Change the Rules for War.* New York: Encounter Books, 2017.

Ramanathan, K. "An Overview of Technology Transfer and Technology Transfer Models." Conference paper presented at the International Conference on "South–South Cooperation for Technology Transfer and Development of Small and Medium Enterprises (SMEs)," Colombo, Sri Lanka, 2008. https://tto.boun.edu.tr/.

———. "The Role of Technology Transfer Services in Technology Capacity Building and Enhancing the Competitiveness of SMEs." Conference paper presented at the Mongolia National Workshop on "Subnational Innovation systems and Technology Capacity-building Policies to Enhance Competitiveness of SMEs," UN-ESCAP and ITMRC, Ulaanbaatar, Mongolia, March 22, 2007.

Ratzel, Friedrich. "As Leis do Crescimento Espacial dos Estados" [The laws of state spatial growth]. In *Ratzel: geografia*, by Antonio C. Robert de Moraes. São Paulo: Ática, 1990. https://repositorio.usp.br/.

———. *La Géographie politique: Les concepts fondamentaux* [Political geography: Fundamental concepts]. Paris: Fayard, 1987.

Ribeiro, Cássio Garcia, and Edmundo Inácio Júnior. "Política de offset em compras governamentais: uma análise exploratória" [Offset policy in government purchases: An exploratory analysis]. Discussion text. Brasília: Instituto de Pesquisa Econômica e Aplicada, n.d.

Ricupero, Rubens. *A diplomacia na construção do Brasil: 1750 – 2016* [Diplomacy in the construction of Brazil: 1750–2016]. Rio de Janeiro: Versal, 2017.

Roberts, James Q. "Maskirovka 2.0: Hybrid Threat, Hybrid Response." Paper. JSOU Occasional Paper. Joint Special Operations University, December 2015. https://apps.dtic.mil/.

Rodrigues, Lysias A. *Geopolítica Do Brasil* [Brazil's geopolitics]. Rio de Janeiro: Biblioteca Militar, 1947.

Rosen, Len. "No One Should Think That Money Spent on NASA Is a Waste." *21st Century Tech Blog (blog)*, September 16, 2014. https://www.21stcentech.com/.

Royal Canadian Air Force Command. B-GA-402-003/FP-001 Royal Canadian Air Force Doctrine. *Force Sustainment*. Canadian Forces Aerospace Warfare Centre, July 2017.

Sá, Fernando Moreira de. "Crónicas do Rochedo XIV: Uma direita musculada numa Espanha dividida" [Chronicles of the Rock XIV: a muscular right in a divided Spain]. Aventar, October 13, 2017. https://aventar.eu/.

Saab. "Gripen E - Sempre à Frente" [Gripen E: Always ahead]. Saab (website), May 30, 2020. https://web.archive.org/web/20200926151350/ https://saab.com/pt/region/brasil/gripenbrasileiro/sempreafrente/.

Sack, Robert David. *Human Territoriality: Its Theory and History*. Cambridge: Cambridge University Press, 1986.

Sanches, Luiz Antonio Ugeda. "Geolaw and the Geographic-Cartographic Construction as Instrument of Public Politics in the Electric Sector." *Electronic Magazine: Time - Technique - Territory* 5, no. 2 (2014): 55–76. https://doi.org/10.26512/ciga.v5i2.15401.

Santos, Aline Oliveira dos, Graziele Priscila Menezes Neiva, Rosiane dos Santos Silva, Célia de Lima Pizolato, and Luciana Maria Gasparelo Spigolon. "Conceito cidade aeroporto: Guarulhos como aerotropolis." [Airport city concept: Guarulhos as aerotropolis]. Seminar paper presented at the X FATECLOG - Logística 4.0 & a Sociedade do Conhecimento, Guarulhos, São Paulo, 2019. https://fateclog.com.br/.

Santos, Carlos Roberto. "O Programa Binacional A-Darter" [The Binational A-Darter Program]. Slide presentation used in a lecture at GAC AFS headquarters, on the visit of Minister of State for Foreign Affairs of Brazil Ambassador Mauro Vieira and his delegation, March 31, 2015. Head of the A-Darter Project Monitoring and Control Group in South Africa –GAC AFS. Centurion, South Africa.

———. "O projeto binacional A-Darter e o seu modelo de transferência da tecnologia" [The binational A-Darter project and its technology transfer model]. Master's dissertation, Universidade da Força Aérea, Rio de Janeiro, 2019.

Santos, Cristiane dos. "Prontos a Qualquer Hora" [Ready anytime]. *Aerovisão - Revista Da Força Aérea Brasileira*, 2019.

Santos, Cristiane dos, and Jonathan Jayme. "Operação Artemis: FAB Em Conflito Real" [Operation Artemis: FAB in real conflict]. *Aerovisão - Revista Da Força Aérea Brasileira*, 2018.

Santos, Jean Carlos Silva dos. "Strategic Information Management as a Conditioning Factor for the Implementation of Defense and National Security Policies in the Context of the Legal Amazon: SIPAM/SIVAM case." Master's dissertation, Getúlio Vargas Foundation (FGV), Rio de Janeiro, 2007.

Santos, Milton. *A Natureza do Espaço: Técnica e Tempo, Razão e Emoção* [The nature of space: Technique and time, reason and emotion]. 4th ed. São Paulo: Edusp, 2014.

———. *Metamorfoses do Espaço Habitado: Fundamentos Teóricos e Metodológicos da Geografia* [Metamorphoses of inhabited space: Theoretical and methodological foundations of geography]. 5th ed. São Paulo: Hucitec, 1997.

Santos, Mizaele. "Inteligência cultural: uma breve análise sobre suas influências à mesa das negociações" [Cultural intelligence: A brief analysis of its influences at the negotiations table]. *C@LEA - Cadernos de Aulas do LEA*, no. 7 (2018): 132–40.

Santos, Murillo. *A Evolução Do Poder Aéreo* [The evolution of airpower]. Rio de Janeiro: Itatiaia, 1989.

Santos, Soraia Emilia Amin dos. "A atuação da Força Aérea Brasileira na operação COVID-19: um estudo de caso" [The performance of the Brazilian Air Force in the COVID-19 operation: a case study]. Thesis. Portugal Military University Institute, Department of Graduate Studies, Joint Staff Course 2021/2022. https://comum.rcaap.pt/.

Schmitt, Michael N. "International Law and Military Operations in Space." In *Max Planck Yearbook of United Nations Law*, vol. 10, edited by A. von Bogdandy and R. Wolfrum, 89–125. Leiden, The Netherlands: Brill, 2006. https://www.mpil.de/.

Schuurman, Bart. "Clausewitz and the 'New Wars' Scholars." *The US Army War College Quarterly: Parameters* 40, no. 1 (March 1, 2010). https://doi.org/10.55540/0031-1723.2515.

Schwartz, Norton A., and Robert B. Stephan. "Don't Go Downtown without Us: The Role of Aerospace Power in Joint Urban Operations." *Aerospace Power Journal*, 2000, 3–11.

Schwartz, Peter. *Cenários: As Surpresas Inevitaveis* [Scenarios: The inevitable surprises]. Rio de Janeiro: Campus, 2003. https://www.academia.edu/.

Selding, Peter B. "Former Arianespace Chief Says SpaceX Has Advantage on Cost." SpaceNews, March 18, 2014. https://spacenews.com/.

Sheehan, Michael. *The International Politics of Space.* 1st ed. New York: Routledge, 2007.

Silva, Amanda Carolina Moura, Gabriel Figueiredo e Lorenzo, Herbert de Souza Pereira, Isabela Carvalho Mustafa Tanajura, Victória Costa Veiga, and Matheus de Oliveira Souza. "Cooperação Brasil-Africa do Sul na produção do míssil A-Darter e a efetivação da estratégia nacional de defesa" [Brazil–South Africa cooperation in the production of the A-Darter missile and the effectiveness of the national defense strategy]. Conference paper presented at the XVI Congresso Acadêmico sobre Defesa Nacional, Rio de Janeiro, 2019.

Silva, Luiz Fernando Póvoas da. "Apresentação Para o Programa de Ciências Aeroespaciais Da UNIFA" [Presentation for the UNIFA aerospace sciences program]. Seminar presentation (paper written in 2010), Rio de Janeiro, May 2018.

Silva, Maria Célia Reis Barbosa da. "Interlacement of culture and defence in the United States and Brazil." In *Culture and Defence in Brazil: An Inside Look at Brazil's Aerospace Strategies,* 1st ed., edited by Maria Filomena Fontes Rico. London: Routledge, 2017.

Silva, Peterson Ferreira da. "A cooperação internacional em programas de produtos de defesa e seus atores: o caso Brasil e África do Sul no desenvolvimento do míssil A-Darter" [International cooperation in defense product programs and its actors: the case of Brazil and South Africa in the development of the A-Darter missile]. Master's thesis, Universidade Estadual de Campinas, 2011. https://repositorio.unesp.br/.

———. "A política industrial de defesa no Brasil (1999–2014): intersetorialidade e dinâmica de seus principais atores" [Defense industrial policy in Brazil (1999–2014): intersectoriality and dynamics of its main actors]. Dissertation. Instituto de Relações Internacionais da Universidade de São Paulo, 2015.

Smith, William. "Asia." *Dictionary of Greek and Roman Geography.* Boston: Little, Brown and Company, 1870. http://www.perseus.tufts.edu/.

Stephens, Dale. "The International Legal Implications of Military Space Operations: Examining the Interplay between International Humanitarian Law and the Outer Space Legal Regime." *International Law Studies* 94 (2018): 75–101.

Teixeira Júnior, Augusto W. M., and Maria Eduarda L. S. Freire. "A importância da Interoperabilidade como Instrumento de Convergência nas Operações Militares do Brasil" [The importance of interoperability as an instrument of convergence in military operations in Brazil]. *Centro de Estudos Estratégicos do Exército: Artigos Estratégicos* 6, no. 1 (June 27, 2019): 29–42.

Teng, Jimmy. *Musket, Map and Money: How Military Technology Shaped Geopolitics and Economics.* 1st ed. Warsaw: De Gruyter Open Poland, 2014.

Thomas, David, Yuan Liao, Zeynep Aycan, Jean-Luc Cerdin, Andre Pekerti, Elizabeth Ravlin, Günter Stahl, et al. "Cultural Intelligence: A Theory-Based, Short Form Measure." *Journal of International Business Studies* 46 (January 29, 2015): 1099–1118. https://doi.org/10.1057/jibs.2014.67.

Thomas, David C., and Kerr Inkson. *Inteligência Cultural: Instrumentos para Negócios Globais* [Cultural intelligence: Tools for global business]. Rio de Janeiro: Record, 2006.

Thompson, Julian. *The Lifeblood of War: Logistics in Armed Conflict.* English ed. London: Potomac Books Inc., 1991.

Tilly, Charles. *Coerção, Capital e Estados Europeus. 1990-1992.* [Coercion, capital and European states. 1990–1992]. 1st ed. São Paulo: EDUSP, 1996.

Tuttle, William, Jr. *Defense Logistics for the 21st Century.* Annapolis, MD: Naval Institute Press, 2013.

United Nations. "Brazil." UN.org. Accessed November 28, 2022. https://www.un.org/.

———. "Charter of the United Nations and Statute of the International Court of Justice." San Francisco: Codification Division Publication, 1945. https://treaties.un.org/.

———. "Guidelines: Peacekeeping Capability Readiness System (PCRS)." United Nations Department of Peace Operations, Department of Operational Support, 2019. https://pcrs.un.org/.

———. "Security Council Presidency." UN.org. Accessed November 28, 2022. https://www.un.org/.

———. *United Nations Peacekeeping Missions Military Aviation Unit Manual.* Department of Peace Operations of United Nations, 2015. http://www.enopu.edu.uy/.

United Nations Commission on International Trade Law (UNCITRAL). *UNCITRAL Legal Guide on International Countertrade Transactions.* New York: United Nations, 1993. https://uncitral.un.org/.

United Nations Office for Outer Space Affairs (UNOOSA). "Space Law Treaties and Principles." UNOOSA.org. Accessed December 20, 2022. https://www.unoosa.org/.

Vasconcellos, Iris. "Com balanço positivo, KC-390 conclui participação no Exercício Culminating" [With positive balance, KC-390 concludes participation in the Culminating Exercise]. Brazilian Air Force (website). Accessed May 2, 2022. https://www.fab.mil.br/.

Vianna, Eduardo Wallier, and Jose Ricardo Souza Camelo. "Defesa Cibernética no Brasil: Primícias de uma História de Sucesso" [Cyber Defense in Brazil: First Friends of a Success Story]. *Revista da Escola Superior de Guerra* 35, no. 75 (set./dec. 2000): 127–54. https://revista.esg.br/.

Vidal, Ludovic-Alexandre, Franck Marle, and Jean-Claude Bocquet. "Measuring Project Complexity Using the Analytic Hierarchy Process." *International Journal of Project Management* 29, no. 6 (2011): 718–27. https://doi.org/10.1016/j.ijproman.2010.07.005.

Viotti, Eduardo B. "Ciência e tecnologia para o desenvolvimento sustentável Brasileiro" [Science and technology for Brazilian sustainable development]. In *Ciência, Ética e Sustentabilidade: Desafios Ao Novo Século*. São Paulo: Cortez; UNESCO, 2001.

Visacro, Alessandro. *A guerra na era da informação* [War in the information age]. *Editora Contexto* (blog), June 8, 2018. https://blog.editoracontexto.com.br/.

———. "Inteligência cultural: assunto impositivo na formação do militar moderno e fundamental no estudo de situação: uma abordagem da temática indígena na Amazônia" [Cultural intelligence - imposing subject in training the modern military and fundamental in the study of the situation: An approach to the indigenous theme in the Amazon]. *Coleção Meira Mattos: revista das ciências militares*, no. 25 (July 14, 2012). http://www.ebrevistas.eb.mil.br/.

Vitale, Michael C. "Jointness by Design, Not Accident." *Joint Force Quarterly*, 1995, 24–30.

Vlasic, Ivan A. "Disarmament Decade, Outer Space and International Law." *McGill Law Journal* 26, no. 2 (September 1980): 135–206. https://lawjournal.mcgill.ca/.

Wallerstein, Immanuel Maurice. *Geopolitics and Geoculture: Essays on the Changing World-System*. Cambridge: Cambridge University Press, 1991.

Waltz, Kenneth N. *Theory of International Politics*. Long Grove, IL: Waveland Press, 2010.

Wanderley, Nelson Freire Lavenère. *Historia Da Força Aerea Brasileira* [History of the Brazilian Air Force]. Brasília: Brazilian Air Force Ministry, 1957.

Weizman, Eyal. "The Politics of Verticality." openDemocracy.net, April 23, 2002. https://www.opendemocracy.net/.

White, Bret. "Reordering the Law for a China World Order: China's Legal Warfare Strategy in Outer Space and Cyberspace." *Journal of National Security Law & Policy* 11, no. 2 (2021): 435–87. https://jnslp.com/.

Wilkerson, Lawrence B. "What Exactly Is Jointness?" *Joint Force Quarterly (Summer 1997)*, 66–68. https://ndupress.ndu.edu/.

Williams, Alison J. "A Crisis in Aerial Sovereignty? Considering the Implication of Recent Military Violations of National Airspace." *Area 42*, no. 1 (March 2010): 51–59.

———. "Hakumat al Tayarrat: The Role of Air Power in the Enforcement of Iraq's Boundaries." *Geopolitics* 12, no. 3 (July 31, 2007): 505–28. https://doi.org/10.1080/14650040701305690.

World Bank Group. *Air Transport: Annual Report 2019*. Washington, DC: World Bank Group, 2020. https://documents1.worldbank.org/.

World Health Organization. "Coronavirus disease (COVID-19) Pandemic." https://www.who.int/.

Wortmeyer, Daniela. "Introdução a Uma Perspectiva Psicossociológica Para o Estudo Das Forças Armadas" [Introduction to a psychosociological perspective for the study of the armed forces]. In *Defesa, Segurança Internacional e Forças Armadas*, edited by Eduardo Svartman, Maria Celina d'Araujo, and Samuel Alves Soares. Campinas, SP, Brazil: Mercado das Letras, 2009.

Brazilian Governmental Sources

Aerospace Museum. "Dia da aviação de asas rotativas" [Rotowing aviation day]. Brazilian air force Historico-Cultural Institute of Aeronautics. Accessed September 21, 2022. https://www2.fab.mil.br/.

Air Force. "Notícias Sobre 'A-DARTER' " [News about 'A-Darter']. n.d. https://www.fab.mil.br/.

Air Force Agency. "EXOP Tápio totaliza cerca de 1.200 horas de voo durante treinamento em Campo Grande" [EXOP Tápio totals around 1,200 flight hours during training in Campo Grande]. Defense News Portal. DefesaNet, May 20, 2019. https://www.defesanet.com.br/.

———. "A CARSAMMA." Accessed May 19, 2022. http://portal.cgna.decea.mil.br/.

Air Force Chief of Staff. DMA 1-1 *Doutrina Básica Da Força Aérea Brasileira* [Basic doctrine of the Brazilian Air Force]. Brazilian Ministry of Aeronautics, July 7, 1975.

———. "Exercício simula cenários de guerra e treina militares para situações reais" [Exercise simulates war scenarios and trains soldiers for real situations]. Força Aérea Brasileira. Accessed May 2, 2022. https://www.fab.mil.br/.

Air Force Chief of Staff and Air Force Command. "Decree No. 393/GC4." *Boletim do Comando da Aeronáutica* 49 (March 24, 2020).

Air Force Command. DCA 2-1/2003 *Doutrina de Logística da Aeronáutica* [Doctrine of aeronautical logistics]. Brazilian Air Force Command, 2003. https://www.sislaer.fab.mil.br/.

———. *PCA 11-47 PLANO ESTRATÉGICO MILITAR DA AERONÁUTICA 2018 - 2027* [Military strategic plan for aeronautics 2018–2027]. Air Force Command, December 20, 2018. https://www.fab.mil.br/.

———. *Relatório Final de Exercício: ExOp Tápio II* [Final exercise report: ExOp Tápio II]. Classified Exercise Report. Brasília: Air Force Command, 2019.

Air Force Command and Air Force Chief of Staff. "Contrato 001/DCTA-SDDP/2006" [Contract 001/DCTA-SDDP/2006]. Brazilian Ministry of Defense, 2006.

Air Operations General Command. "DCAR 100B Capacitação de Recursos Humanos No Âmbito Do COMGAR" [DCAR 100B Training of human resources within the scope of COMGAR]. Brazilian Air Force Command, 2015.

Brazilian National Congress. Decreto Legislativo No 373, de 25 de Setembro de 2013. https://www2.camara.leg.br/.

———. Legislative Decree 373, Pub. L. No. 373 (July 17, 2013). https://www2.camara.leg.br/.

———. National Defense Policy and National Defense Strategy. Brazilian National Congress, 2005. http://www.planalto.gov.br/.

———. National Defense Policy and National Defense Strategy. Brazilian National Congress, 2012. https://www.gov.br/.

———. National Defense Policy and National Defense Strategy. Brazilian National Congress, 2020. https://www.gov.br/.

———. National Defense Policy and National Defense Strategy (Draft). Brazilian National Congress, 2016.

Chamber of Deputies. Decree 5484, Pub. L. No. 5484 (June 30, 2005). https://www2.camara.leg.br/.
———. Supplementary Law 97, Pub. L. No. 97 (June 9, 1999). https://www2.camara.leg.br/.
Civil Office. Decree 4200, Pub. L. No. 4200 (April 17, 2002). http://www.planalto.gov.br/.
———. Decree 4240, Pub. L. No. 4240 (May 21, 2002). http://www.planalto.gov.br/.
———. Decree 6617, Acordo entre o Governo da República Federativa do Brasil e o Governo da República da África do Sul no Campo da Cooperação Científica e Tecnológica [Agreement between the Government of the Federative Republic of Brazil and the Government of the Republic of South Africa in the Field of Scientific and Technological Cooperation] (October 23, 2008). http://www.planalto.gov.br/.
———. Decree 6834 (April 30, 2009). http://www.planalto.gov.br/. Accessed December 15, 2022.
———. Decree 8122, Pub. L. No. 8122 (October 16, 2013). http://www.planalto.gov.br/.
———. Decree 7970 (March 28, 2013). http://www.planalto.gov.br/.
———. Decree 12598, Pub. L. No. 12598 (March 21, 2012). http://www.planalto.gov.br/.
———. Decree 73,160, November 14, 1973. https://www2.camara.leg.br/. Accessed December 14, 2022.
———. Decree of October 18, 1999 (Repealed by Decree no. 9829 of 2019). http://www.planalto.gov.br/.
———. Federal Constitution, § Article 7 (1967). http://www.planalto.gov.br/.
Federal Audit Court. *Audit Report: Implementation of the SIVAM Project*. Period from June 1, 2002, to October 31, 2002. Judgment 194/2003 – Plenary Session. Accessed December 14, 2022. https://www.jusbrasil.com.br/.
Ministry of Aeronautics. "Decree 1000 - GM2 and 'Doutrina Básica Da Força Aérea'" [Decree 1000-GM2 and basic air force doctrine]. *Boletim Do Ministério Da Aeronáutica*, April 30, 1959, 4th ed.
Ministry of Defense (MOD). "Acórdão 194/2003—Plenário, Doc AC-0194-07/03-P" [Decision 194/2003-Plenary, Doc AC-0194-07/03-P]. 2003.
———. *Apêndice V do contrato 001/DCTA-SDDP/2006, Transferência de tecnologia do Projeto A-Darter* [Appendix V of contract 001/

DCTA-SDDP/2006, Transfer of technology from the A-Darter Project]. Brazilian Ministry of Defense, 2006.

———. "Censipam Institutional Folder." SIPAM, 2015. www.sipam.gov.br.

———. DCA 1-1, *DOUTRINA BÁSICA DA FORÇA AÉREA BRASILEIRA* [DCA 1-1 basic doctrine of the Brazilian air force]. Brazilian Ministry of Defense, 2020. https://www2.fab.mil.br/.

———. *DOUTRINA DE OPERAÇÕES CONJUNTAS 2Vol* [Joint operations doctrine, vol. 2]. Brazilian Chairman of the Joint Chiefs of Staff (EMCFA), 2020. https://www.gov.br/.

———. "Glossário das Forças Armadas MD35-G-01" [Glossary of the armed forces MD35-G-01]. Brazilian Chairman of the Joint Chiefs of Staff (EMCFA), 2015. https://bdex.eb.mil.br/.

———. "ICA 100-40 AERONAVES NÃO TRIPULADAS E O ACESSO AO ESPAÇO AÉREO BRASILEIRO" [ICA 100-40 Unmanned aircraft and access to Brazilian airspace]. Brazilian Ministry of Defense, 2020. https://www.gov.br/.

———. "Lei Complementar No 97, de 9 de Junho de 1999" [Complementary law no. 97, of June 9, 1999]. Brazilian Ministry of Defense, June 9, 1999. http://www.planalto.gov.br/

———. *Livro Branco de Defesa Nacional [Defense white paper]*. Brazilian National Congress, 2020. https://www.gov.br/.

———. "Ministerial Directive No. 003/2002 - Military Commands and the Defense Chief of Staff's Participation in the Activation of SIPAM," March 4, 2002.

———. "Ordem Do Dia Alusiva Ao 31 de Março de 1964" [Agenda allusive to March 31, 1964]. Brazilian Ministry of Defense, March 31, 2021. https://www.fab.mil.br/.

———. "Política Nacional de Defesa e Estratégia Nacional de Defesa" [National defense policy and national defense strategy]. Brasília: Ministério da Defesa, Secretaria de Assuntos Estratégicos, 2018. https://www.gov.br/.

———. "Portaria Normativa n.o 61/GM-MD, de 23 de Outubro de 2018" [Normative Ordinance No. 61/GM-MD, of October 23, 2018]. Brazilian Ministry of Defense, 2018.

———. Regulatory Ordinance 61/GM-MD, Pub. L. No. 61/GM-MD (2018). https://www.in.gov.br/.

———. "Revisão Do Planejamento Estratégico Da CENSIPAM" [Review of CENSIPAM's strategic planning]. General Secretary of Brazilian Ministry of Defense, 2018. https://www.gov.br/.

———. "Termos Aditivos ao contrato com a ARMSCOR." Addendum to contract 001/CTA-SDPP/2006. Brazilian Ministry of Defense, 2011.

Ministry of Defense and Air Force Command. "Concepção Estratégica - Força Aérea 100" [Strategic design—Air Force 100]. Brazilian Ministry of Defense, 2018. https://www.fab.mil.br/.

———. Decree 126/SDAD, Pub. L. No. 126/SDAD (December 16, 2019).

———. Decree 1008/GC3, Pub. L. No. 1008/GC3 (2020). https://pesquisa.in.gov.br/.

———. Decree 1547/GC3, Pub. L. No. 1547/GC3 (2018). https://www.in.gov.br/.

———. Decree 2102/GC3, December 18, 2018.

———. Decree 2146/GC3 (2018). https://www.in.gov.br/.

———. Ordinance 1395/GC4, Pub. L. No. 1395/GC4 (2005). https://www.jusbrasil.com.br/.

Ministry of Defense and Joint Chiefs of Staff. *Doutrina militar de defesa cibernética* [Cyber defense military doctrine]. Ministério da Defesa, 2014.

———. MD34-M-03, *Manual de Emprego Do Direito Internacional Dos Conflitos Armados (DICA) Nas Forças Armadas* [MD34-M-03, Manual for the use of international law on armed conflicts in the armed forces]. 2011. https://www.gov.br/.

———. MD42-M-02 *Doutrina de Logística Militar* [MD42-M-02 Military logistics doctrine]. Brazilian Ministry of Defense, 2016. https://www.gov.br/.

Ministry of Defense and Secretariat for Strategic Affairs of the Presidency of the Republic. Decree no. 6703, Approves the National Defense Strategy, and takes other measures. December 18, 2008. http://www.planalto.gov.br/.

Ministry of Foreign Affairs. "República da África do Sul" [Republic of South Africa]. Brazilian Ministry of Foreign Affairs (website), 2010.

Ministry of Justice and Public Security. *FUNAI: Autonomy and Indigenous Protagonism*. Translated by Guilherme Lucas Rodrigues Monteiro. Brasilia: Ministry of Justice and Public Security, 2022. https://www.gov.br/.

Ministry of Justice, Ministry of Aeronautics, and Secretariat of Strategic Affairs. Exposição de Motivos N°194 [Explanatory memorandum no. 194]. *Diário Oficial Da União*. September 24, 1990, sec. I.

Ministry of Science and Technology. Acordo entre o Governo da República Federativa do Brasil e o Governo da República da África do Sul no Campo da Cooperação Científica e Tecnológica.

[Agreement between the Government of the Federative Republic of Brazil and the Government of the Republic of South Africa in the Field of Scientific and Technological Cooperation]. http://www.planalto.gov.br/.

Ministry of Transport. *Anuário Estatístico de Transportes 2010-2018* [Statistical Yearbook of Transport 2010-2018]. Brazil's Ministry of Transport, 2019. https://www.gov.br/.

National Civil Aviation Agency. "Requisitos Gerais para Aeronaves não-tripuladas de Uso Civil - Regulamento Brasileiro da Aviação Civil Especial" [General requirements for civilian unmanned aircraft–Brazilian special civil aviation regulation]. RBAC-E n°94. Brazilian National Civil Aviation Agency, 2017. https://www.anac.gov.br/.

United States of Brazil. Constitution of the United States of Brazil (February 24, 1891), § Article 88 (1891). http://www.planalto.gov.br/.

———. Constitution of the United States of Brazil (July 16, 1934), § Article 4 (1934). http://www.planalto.gov.br/.

———. Constitution of the United States of Brazil (September 18, 1946), § Article 4 (1946). http://www.planalto.gov.br/.

Index

aereality, 122
aeronautics: 29–31, 33, 49, 50, 123, 132, 175; Ministry of, 29, 31
Aeronautics Command, 30, 33, 49
aerospace: environment, xviii, 4, 37, 61, 116–19, 125, 126, 194; geopolitics, xviii, 59–61, 115, 117, 122–26, 194; industry, xiii, xv, xix, 47, 61, 103, 117, 121, 122, 124, 131, 132, 161, 194, 195
Aerospace Command, 35
air assets, 86, 90
air defense, xiv, 4, 33–35, 43, 51, 54
Air Force Command, 47, 48, 50, 81, 89, 139, 169, 196
air traffic control, xiv, xvi, xvii, 3, 4, 7, 33–35, 44, 50, 51, 54, 193
air traffic management, xvii, 3, 4, 33–35, 43–45, 47, 48, 51, 52, 193
air transportation, 121, 123, 125, 126
air unit, 79, 85, 86, 88–90
airborne crew, 82, 87, 94, 97, 99, 101n31
airplane, 116, 119
airpower, xiii, xv, xx, 10, 70, 72, 79, 81, 122
airspace, xiv, xv, xvii, xviii, 4, 12, 27, 31, 33, 35, 37, 39, 40, 43–49, 52–54, 61, 81, 89, 95, 116, 119, 120, 124–26
air-to-air missile, 180
Amazon: xvi, 4, 23, 27–32, 36–40, 43, 98, 107; Amazon Protection System, xvi, 4, 27, 29, 30; Amazon Surveillance System, xvi, 4, 27, 29–31
asteroid, 61, 120
Avibras, 176, 177, 179, 181, 184, 186, 187
axiom, 136–38, 143, 144, 148, 149
Bozeman Model, 170, 172, 174
Brazilian Air Force Cyber Defense Center, 3, 20, 21
Brazilian defense industry, 156, 158, 170, 177, 188
Brazil–South Africa partnership, xx, 132, 174–77, 179, 180, 183, 189, 190

civilian, xvi, 3, 7, 23, 34, 39, 66, 70–73, 122, 155, 166, 178
combat support, 131, 145, 149
complex adaptive system, 10
Contingent Effectiveness Model, 170, 172, 183, 189
cyber defense, xi, xvi, 3, 19–22, 24, 71, 73, 74, 193
cyber strategic sector, 19
Defense: Minister of, 36, 37, 154, 176; Ministry of, 12, 19, 30, 32, 46, 67, 69, 73, 88, 93, 155, 158–60, 162, 169
Denel Dynamics, 175, 177–84, 186, 187
dissuasion strategy, 38, 40
drug trafficking, 36–39, 50
economic variable, 120, 121
Embraer EMB-314 Super Tucano A-29, 3, 5, 11–15, 193
Eximbank, 32
factual reference, 135, 142–45, 150
First Generation War, 5
Fourth Generation War, 6, 9, 15
FX-2 Project, 22
geographic variable, 117, 118
geography, xv, 31, 104–6, 115, 116, 118, 119
global stability, xviii, 13, 59, 79–81
Gripen, 22–23, 181, 182, 195
human resources, xiii, 33, 34, 39, 50, 52, 60, 79, 86, 88–90, 117, 135, 161, 171, 193, 195
humanitarian aviation, 59, 79
humanitarian law, xvii, 59, 63–65, 194
hybrid war, xvi, 21, 38, 39
hypothesis of equifinality, 12, 14, 15
ideological variable, 123, 125
ideology, 6, 31, 123, 124
illegal mining, 28, 37–40
infosphere, 4, 38
intelligent weapons, 195
international law, xvii, 59, 63–65, 67–69, 71, 80, 109, 119, 120, 158, 194
international power balance, 8
joint operations, xv, xviii, 8, 60, 71, 83, 90, 93–95, 100, 193, 194

Kármán line, 117
know-how, 172, 174, 185, 186
know-why, 183, 185
material resources, 86, 88, 90, 95, 163
Mectron, 176, 177, 179, 181
militarization, 29, 72, 120, 125, 195
military air unit, 90
military planning, xvi, 7, 8, 10, 13–15, 73, 193
missions, 5, 11, 15, 20, 21, 23, 59, 68, 72, 79, 82–89, 93, 94, 97, 131, 136, 140, 141
multidomain environment, 4, 20, 39
National Defense Policy, xv, 20, 36, 68, 87, 131, 153–56, 169
National Defense Strategy, xv, 18, 19, 36, 38, 68, 81, 93, 154–56
national innovation systems, 171
NK Model, 10, 11, 14, 15
operational risk management, 7, 9, 13, 14, 50
Opto Eletrônica, 177, 179, 181, 186
outer space, xv, xviii, xx, 23, 37, 61, 66, 72, 81, 108, 109, 115–26, 193, 195, 196
peace: xv, xviii, 11, 14, 15, 46, 48, 59, 65, 67–69, 71, 75, 79–90, 93, 96, 108, 194; operations, xv, xviii, 59, 79–82, 84–90, 93, 194
peacekeeping: operations, xvii, xviii, 63, 68, 69, 75, 79, 81, 82, 84, 87, 88; readiness, xviii, 79, 84, 85
political variable, 118–20, 125
prestige, xv, 82, 117, 124, 126
principle of logistics, 148–50
professional training, 79
readiness protocol, 86
representation, 4, 9, 124
Saab, 22, 195
satellite, 23, 71–73, 121, 122, 186
Second Generation War, 6, 15
semantics, 136, 142, 143, 150
simulation, 5, 12, 162, 181, 188
South America, xvii, 36, 38, 43–48, 52, 54, 67, 106, 124, 184
South-South cooperation, 176

sovereignty, xv, xvi, 4, 12, 24, 27–29, 35–37, 39, 61, 64, 80, 107, 108, 111, 119, 120, 125, 153–55, 166, 167
space debris, 109, 118
space race, xv, 120, 122
strategic capabilities, 79
strategic partnership, 169
surveillance, xvi, 4, 7, 23, 27–31, 33, 36–40, 44, 45, 50, 51
technological variable, 122, 123
technology, xiv, xix, xx, 9, 10, 19, 22, 24, 25, 26, 28, 33, 47, 60, 70–72, 74, 116, 117, 153, 155–61, 163–66, 169–79, 183–90, 194; transfer, 22, 132, 164–66, 170–74, 178, 179, 183, 186, 189
telecommunication, 23, 33, 72, 106, 109, 121, 123
territory, xviii, 15, 23, 27, 28, 31, 36, 37, 39, 40, 46, 47, 52, 66, 67, 87, 96, 106, 107, 110, 116, 118–20, 155, 179, 194
Third Generation War, 6
training program, 69, 79
Ukraine, 7, 29, 32
unmanned systems, 7, 217
unmanned vehicles, 6, 70, 193
weaponization, xvi, 5, 7, 13–15, 23, 61, 72, 120, 193